Crop Physiology

BOOKS IN THE BIOTOL SERIES

BIOTECHNOLOGY BY OPEN LEARNING

Crop Physiology

PUBLISHED ON BEHALF OF :

Open universiteit   and   **University of Greenwich (formerly Thames Polytechnic)**

Valkenburgerweg 167
6401 DL Heerlen
Nederland

Avery Hill Road
Eltham, London SE9 2HB
United Kingdom

Butterworth-Heinemann Ltd
Linacre House, Jordan Hill, Oxford OX2 8DP

℞ A member of the Reed Elsevier plc group

OXFORD LONDON BOSTON
MUNICH NEW DELHI SINGAPORE SYDNEY
TOKYO TORONTO WELLINGTON

First published 1994
Reprinted 1995

British Library Cataloguing in Publication Data
A catalogue record for this book is
available from the British Library

Library of Congress Cataloging in Publication Data
A catalog record for this book is
available from the Library of Congress

ISBN 0 7506 0560 X

Composition by University of Greenwich
(formerly Thames Polytechnic)
Printed and Bound in Great Britain by
Martins the Printers Ltd, Berwick-upon-Tweed

The Biotol Project

The BIOTOL team

OPEN UNIVERSITEIT, THE NETHERLANDS
Prof M. C. E. van Dam-Mieras
Prof W. H. de Jeu
Prof J. de Vries

UNIVERSITY OF GREENWICH (FORMERLY THAMES POLYTECHNIC), UK
Prof B. R. Currell
Dr J. W. James
Dr C. K. Leach
Mr R. A. Patmore

This series of books has been developed through a collaboration between the Open universiteit of the Netherlands and University of Greenwich (formerly Thames Polytechnic) to provide a whole library of advanced level flexible learning materials including books, computer and video programmes. The series will be of particular value to those working in the chemical, pharmaceutical, health care, food and drinks, agriculture, and environmental, manufacturing and service industries. These industries will be increasingly faced with training problems as the use of biologically based techniques replaces or enhances chemical ones or indeed allows the development of products previously impossible.

The BIOTOL books may be studied privately, but specifically they provide a cost-effective major resource for in-house company training and are the basis for a wider range of courses (open, distance or traditional) from universities which, with practical and tutorial support, lead to recognised qualifications. There is a developing network of institutions throughout Europe to offer tutorial and practical support and courses based on BIOTOL both for those newly entering the field of biotechnology and for graduates looking for more advanced training. BIOTOL is for any one wishing to know about and use the principles and techniques of modern biotechnology whether they are technicians needing further education, new graduates wishing to extend their knowledge, mature staff faced with changing work or a new career, managers unfamiliar with the new technology or those returning to work after a career break.

Our learning texts, written in an informal and friendly style, embody the best characteristics of both open and distance learning to provide a flexible resource for individuals, training organisations, polytechnics and universities, and professional bodies. The content of each book has been carefully worked out between teachers and industry to lead students through a programme of work so that they may achieve clearly stated learning objectives. There are activities and exercises throughout the books, and self assessment questions that allow students to check their own progress and receive any necessary remedial help.

The books, within the series, are modular allowing students to select their own entry point depending on their knowledge and previous experience. These texts therefore remove the necessity for students to attend institution based lectures at specific times and places, bringing a new freedom to study their chosen subject at the time they need and a pace and place to suit them. This same freedom is highly beneficial to industry since staff can receive training without spending significant periods away from the workplace attending lectures and courses, and without altering work patterns.

BIOcalm

SOFTWARE IN THE BIOTOL SERIES

BIOcalm interactive computer programmes provide experience in decision making in many of the techniques used in Biotechnology. They simulate the practical problems and decisions that need to be addressed in planning, setting up and carrying out research or development experiments and production processes. Each programme has an extensive library including basic concepts, experimental techniques, data and units. Also included with each programme are the relevant BIOTOL books which cover the necessary theoretical background.

The programmes and supporting BIOTOL books are listed below.

Isolation and Growth of Micro-organisms
Book: *In vitro* Cultivation of Micro-organisms
           Energy Sources for Cells

Elucidation and Manipulation of Metabolic Pathways
Books: *In vitro* Cultivation of Micro-organisms
           Energy Sources for Cells

Gene Isolation and Characterisation
Books: Techniques for Engineering Genes
           Strategies for Engineering Organisms

Applications of Genetic Manipulation
Books: Techniques for Engineering Genes
           Strategies for Engineering Organisms

Extraction, Purification and Characterisation of an Enzyme
Books: Analysis of Amino Acids, Proteins and Nucleic Acids
           Techniques used in Bioproduct Analysis

Enzyme Engineering
Books: Principles of Enzymology for Technological Applications
           Molecular Fabric of Cells

Bioprocess Technology
Books: Bioreactor Design and Product Yield
           Product Recovery in Bioprocess Technology
           Bioprocess Technology: Modelling and Transport Phenomena
           Operational Modes of Bioreactors

Further information: Greenwich University Press,
University of Greenwich, Avery Hill Road, London, SE9 2HB.

Contributors

AUTHOR

Dr G.D. Weston, De Montfort University, Leicester, UK

EDITOR

Professor A. Cobb, Nottingham Trent University, Nottingham, UK

SCIENTIFIC AND COURSE ADVISORS

Prof M.C.E. van Dam-Mieras, Open universiteit, Heerlen, The Netherlands

Dr C.K. Leach, De Montfort University, Leicester, UK

ACKNOWLEDGEMENTS

Grateful thanks are extended, not only to the authors, editors and course advisors, but to all those who have contributed to the development and production of this book. They include Ms H. Leather, Mrs A. Liney, Dr S. Shaw, Miss J. Skelton, Dr C.A. Smith, Mrs S. Smith and Professor R. Spier.

The development of this BIOTOL text has been funded by **COMETT, The European Community Action Programme for Education and Training for Technology**. Additional support was received from the Open universiteit of The Netherlands and by University of Greenwich (formerly Thames Polytechnic).

Contents

How to use an open learning text

An open learning text presents to you a very carefully thought out programme of study to achieve stated learning objectives, just as a lecturer does. Rather than just listening to a lecture once, and trying to make notes at the same time, you can with a BIOTOL text study it at your own pace, go back over bits you are unsure about and study wherever you choose. Of great importance are the self assessment questions (SAQs) which challenge your understanding and progress and the responses which provide some help if you have had difficulty. These SAQs are carefully thought out to check that you are indeed achieving the set objectives and therefore are a very important part of your study. Every so often in the text you will find the symbol Π, our open door to learning, which indicates an activity for you to do. You will probably find that this participation is a great help to learning so it is important not to skip it.

Whilst you can, as an open learner, study where and when you want, do try to find a place where you can work without disturbance. Most students aim to study a certain number of hours each day or each weekend. If you decide to study for several hours at once, take short breaks of five to ten minutes regularly as it helps to maintain a higher level of overall concentration.

Before you begin a detailed reading of the text, familiarise yourself with the general layout of the material. Have a look at the contents of the various chapters and flip through the pages to get a general impression of the way the subject is dealt with. Forget the old taboo of not writing in books. There is room for your comments, notes and answers; use it and make the book your own personal study record for future revision and reference.

At intervals you will find a summary and list of objectives. The summary will emphasise the important points covered by the material that you have read and the objectives will give you a check list of the things you should then be able to achieve. There are notes in the left hand margin, to help orientate you and emphasise new and important messages.

BIOTOL will be used by universities, polytechnics and colleges as well as industrial training organisations and professional bodies. The texts will form a basis for flexible courses of all types leading to certificates, diplomas and degrees often through credit accumulation and transfer arrangements. In future there will be additional resources available including videos and computer based training programmes.

Preface

The importance we may apply to the study of plant physiology reflects the dominant roles fulfilled by plants in the biosphere and in the affairs of Man. Plants are the primary gatherers and providers of cellular energy and suppliers of organic nutrients for virtually all other organisms. Some fulfil this role by growing wild in uncultivated areas. Others have been harnessed by Man to provide food and medicines, building materials, clothing and commodities such as paper. Also, of fundamental importance is the growing awareness that increases in the world population and the problems arising from environmentally 'dirty' technologies makes it essential for Mankind to turn to environmentally cleaner production processes. Central to this is the recognition that plants are a renewable resource which may be generated in an environmentally acceptable manner and which could provide a route to achieving an environmentally sustainable development of human society. The attainment of such a laudable objective will only be achieved through improvements in our understanding of how plants function allied to improved strain development and the application of environmentally sensible agricultural practices. The purpose of this text is to provide readers with up-to-date knowledge of plant physiology, especially the physiology of multicellular terrestrial plants.

The two key features of plants are their ability to carry out photosynthesis and their sedentary nature. These two features override the way in which plants organise their physiological functions. In particular, the sedentary nature of plants has many important consequences both on the mechanisms by which they acquire nutrients and the strategies they adopt to respond to their chemical and physical environments. Unlike animals, plants cannot go out in search for food and water. Once their seeds have germinated, plants have no option but to stay put and to make use of the environment in which they find themselves. In general, they have to maximise their contact with their environment in order to gather essential nutrients. Thus, above ground, this contact is maximised by producing an enormous leaf area in order to harvest sunlight and carbon dioxide for photosynthesis. Below ground, the production of an extensive root system facilitate water and mineral uptake. We must anticipate, therefore, a spacial organisation of nutrient uptake and light harvesting within each plant. We must also anticipate that some degree of co-ordination is maintained between these functional parts. We may, therefore, predict that transport systems will be present which enable the transfer of materials from one part of the plant to another. It is also self-evident that the development of each part must be co-ordinated in order to produce plants in which root and shoot capacities are balanced.

In order to understand plant physiology, therefore, we need to have knowledge of the nature of the cells which make up the tissues and organs of plants and of the spacial arrangement of these organs. These are described in Chapter 1 of the text. In Chapter 2 the site and processes of photosynthesis are discussed. In order for photosynthesis to occur, water and essential minerals must be transported from the soil, via roots to the aerial parts of plants. The uptake and transport of water and mineral nutrients are described in Chapters 3 and 4. To complete this picture of transport within plants the movement of the products of photosynthesis (photosynthates) is examined in Chapter 5.

A key point to remember throughout these chapters is that plants are subject to changes in their environment. These may be seasonal or simply diurnal. Water availability, temperature, intensity of incident radiation and space are all subject to changes and each plant must develop mechanisms that can respond to both short-term and longer-term changes. The extent to which each plant can accommodate environmental conditions governs whether it may survive and thrive in a particular niche. In some cases, such responses are short-term and reflect metabolic and structural changes by individual groups of cells. In other cases, the adaptations are more profound. Even two genetically identical seeds germinating in different environments will produce non-identical plants, a reflection of the inherent plasticity of plant development. This plasticity is mediated by phytohormones which not only control development but co-ordinate the development of different plant parts. The structure and activities of the major plant hormones are discussed in Chapter 6 and the reproduction, growth and development of plants are examined in Chapters 7 and 8.

It is impossible to cover all of the issues of plant biology in a single text especially, when one considers the diversity and range of plant types found on Earth. The author's selection of material and his synthesis of a coherent picture of plant biology is excellent and provides readers with opportunities to learn the essential elements of plant physiology. The application of this knowledge to agricultural and horticultural crop products is dealt with in the BIOTOL text, 'Crop Productivity'.

Scientific and Course Advisors: Professor M.C.E. van Dam-Mieras
Dr C.K. Leach

Plant cell structure: variations on a theme

Plant cell structure: variations on a theme

1.1 Introduction

In order to understand how plants work, we must first have some knowledge of their structure. This chapter presents an overview of plant cell structure indicating functions where appropriate.

plant cell
organelles

Plants are multicellular organisms consisting of millions of cells with specialised functions. These cells show wide variations in shape and size but they have a common organisation. At some stage in their life, all plant cells contain a nucleus, cytoplasm and subcellular organelles, all of which are enclosed by a membrane to form what is called the protoplasm, (Figure 1.1.) Examine Figure 1.1 carefully and note the names of the various labelled structures.

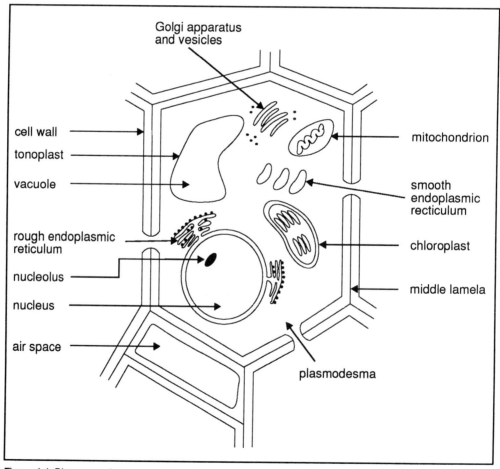

Figure 1.1 Diagrammatic representation of a typical plant cell.

vacuole
In mature cells, the vacuole is extremely large and accounts for more than 90% of total cell volume. It is important in cellular water relations but is also often used as a depository for waste materials. Chloroplasts are the most prominent organelles because of their size and are part of a family of organelles called plastids. Two examples of plastids are chloroplasts, which are green plastids concerned with photosynthesis and amyloplasts, which are colourless plastids concerned with the storage of starch in non-photosynthetic tissues.

plastids

cell wall
The protoplasm is itself surrounded by a cell wall which not only defines the shape of the cell but also provides the basis for distinguishing one cell type from another. It is a striking fact of plant cell structure that some cell types lose their nucleus during development, for example phloem sieve tube cells. Several cells not only lose their nucleus but actually die during their development; examples of these are sclerenchyma, xylem and cork cells. Although the protoplast dies in these cells and is digested and resorbed by adjacent cells, the cell wall persists and the cell still performs a function. This might have contributed to the old idea that plant cell walls are simply dead structures. This idea has, however, given way to a new picture of a highly complex structure performing a variety of functions in the life of the plant. In this chapter we will first examine the generalised structure of a plant to gain an overview. We will follow this by looking at a stylised cell so as to gain an appreciation of general cell structure shown by the fundamental cell types. This chapter is designed to set the scene for the later chapters where we will be examining the functions of these cells and the developmental processes shown by the plant as it completes its life cycle. We will restrict ourselves predominantly to plants which produce seeds.

1.2 An overview of plant structure

Plants show wide diversity of external morphology but they are built on a common basic plan. Figures 1.2a-1.2d show diagrams of a vegetative plant, ie one which is not yet producing flowers. The drawings in these figures have been somewhat generalised. You should anticipate that there are some differences between species. Nevertheless, the general layout of tissues within leaves, stems and roots is rather similar in a wide range of higher plants. We will learn of some of these differences in later sections.

plant organs
Figure 1.2a shows that the plant consists basically of leaves, a stem and a root, each of which is referred to as an organ. The stem and the leaves together are called the shoot. The position of attachment of each leaf to the stem is called the node and the portion of the stem between nodes is an internode. Internodes can be long as in runner or stick beans or short as in bush or dwarf beans. The stem has a bud at its apex called the apical bud which contains a zone of production of new cells. Such zones are called meristems and that in the apical bud is called the apical meristem. The action of this meristem is to produce new leaves, nodes and internodes and thereby to control the growth of the shoot. Leaves carry a bud in the upper angle between themselves and the stem, (the leaf axil). These buds are called axillary buds and are, in essence, replicas of the apical bud. They are responsible for producing the side branches of the plant. However, they normally go through a period of inactivity of several weeks or so before they start to grow out into side branches. Because the duration of this inactivity is governed by the apical bud, the phenomenon is called apical dominance. Species vary regarding the relative degree or strength of apical dominance and this has an effect on the overall shape of the shoot.

meristems

apical dominance

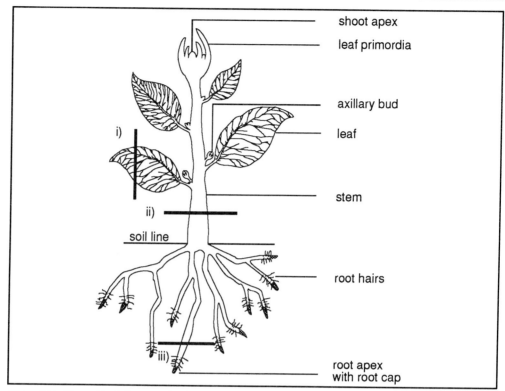

Figure 1.2a Generalised diagram of a plant. i), ii) and iii) refer to sections shown in Figures 1.2b, 1.2c and 1.2d.

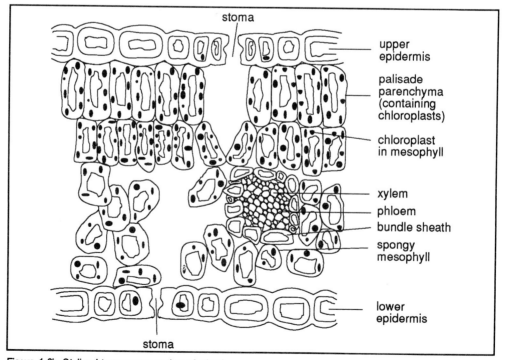

Figure 1.2b Stylised transverse section of a leaf.

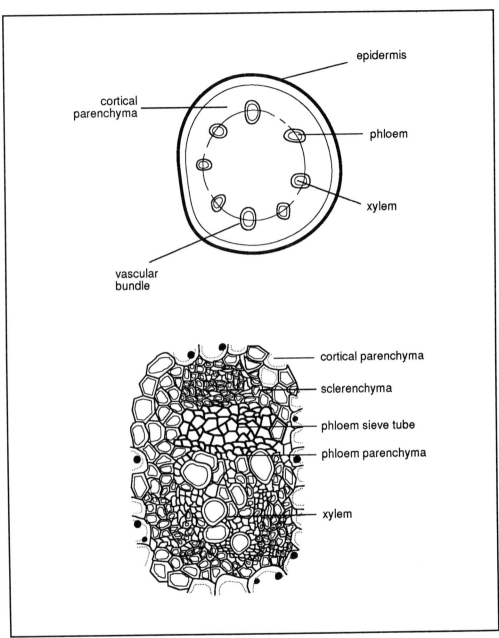

Figure 1.2c Stylised transverse section of a stem with a detail of a vascular bundle.

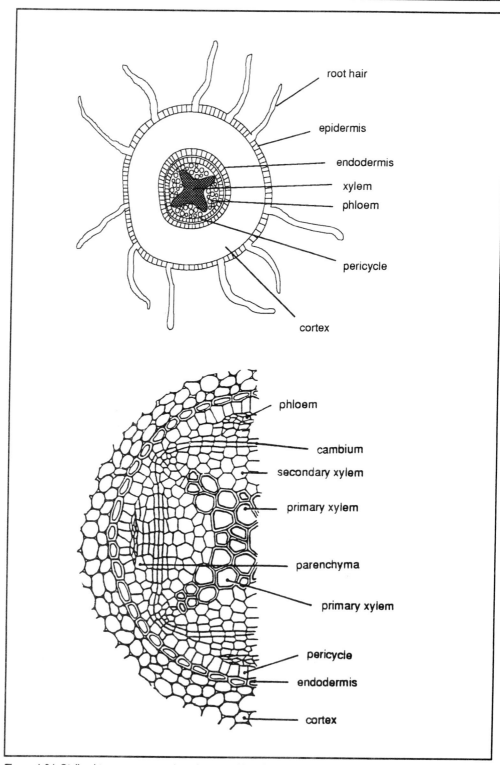

Figure 1.2d Stylised transverase section of a root with a detail of the vascular tissue.

SAQ 1.1

Look at the two diagrams below which are designed to show the shape of a shoot with strong or weak apical dominance. Decide which is which, giving brief reasons for your conclusion.

a)   b)

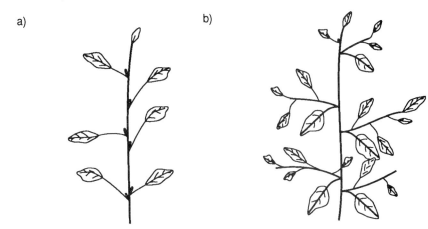

Damage to the shoot apex, as occurs in browsing by animals, results in the release of apical dominance and the outgrowth of previously inactive lateral buds.

root growth As you know, the root grows down into the soil and one of its roles is that of anchorage. It is aided in this by the production of branch or lateral roots. The growth of the main root is governed by a meristem at its apex and replicas of this are produced a few centimetres away from the apex which grow out to produce the lateral roots. Lateral roots are never produced close to the apical meristem and there is a phenomenon in roots analogous to apical dominance in shoots. Two extremes of it produce on the one hand, a tap root system which consists of a major root with very few subsidiary lateral roots, and on the other hand a fibrous root system in which numerous laterals grow as actively as the main root. Indeed, in fibrous root systems it is often impossible to decide which is the main root.

Figure 1.2 also shows the internal arrangement of tissues in leaves, stems and roots and descriptions of the cells which make up these tissues will be described later in the chapter. Before proceeding with this, it is opportune to point out that the morphology and anatomy referred to is that of plants which belong to the family Dicotyledonae, known as dicotyledons or dicots. The second major family is the Monocotyledonae,
dicots known as monocotyledons or monocots. Monocots contain the same cells and tissues
monocots as dicots but there are a few general differences. Monocot leaves have parallel veins and the leaves are often long and narrow, as in wheat and maize. Stems possess vascular bundles as in dicots, but they are usually spread throughout the cross section of the stem rather than in a single ring. Monocot roots differ from dicot roots in having a central pith, with xylem and phloem in bundles arranged radially.

1.3 Cell structure

1.3.1 The wall is formed of layers

In a later chapter we will be examining the detailed structure of cell walls to try to explain what happens when plant cells grow. It would be useful, however, to know something about the basic structure, for this will help your understanding of the section which follows. The terms pectic substances, hemicellulose and cellulose will be introduced here but the nature of the material will not be described until a later chapter. The general arrangement of plant cell walls is illustrated in Figure 1.3: note its layered structure.

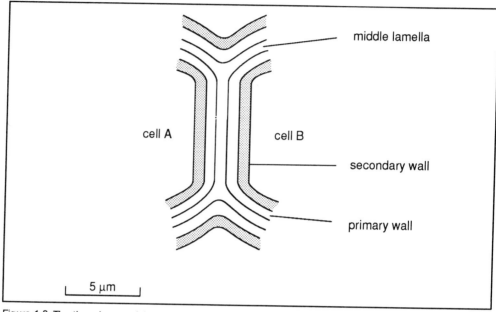

Figure 1.3 The three layers of the cell wall. The bar is 5µm. In growing cells only the primary wall and middle lamella are present. The secondary wall, which can be very thick, is added when growth ceases.

middle lamella

The middle lamella is the material which holds together two adjacent cells and is made predominantly of pectic substances. The cell wall of a growing cell possesses another layer, the primary wall, which contains small amounts of pectic substances but mainly consists of hemicellulose and cellulose. A small number of cell types possess no other layers in their walls but most contain a third layer, called the secondary wall, which is laid down when the cell stops growing. It contains hemicellulose and cellulose but no pectic substances. The pectic substances, hemicellulose and cellulose have in common the fact that they are sugar polymers of various sizes. Cellulose is the most highly polymerised and it forms fibres which are the major structural entity of the wall. In most cell walls, the spaces between the polymers are occupied by water and small molecular weight compounds can diffuse freely between the polymers. In other cells the wall is impregnated with one of two substances, suberin or lignin, which fill the spaces between the polymers. These two materials are hydrophobic and they effectively waterproof the cell wall where they are laid down. Cork cells are an example of cells with suberin-impregnated walls and it is the suberised walls which give cork its resilience. As we shall see later in this chapter, xylem vessels, tracheids and sclerenchyma cells have lignin-impregnated walls. Lignin rigidifies and strengthens walls and the completed product with cellulose fibres embedded in a matrix of lignin is analogous to steel-reinforced concrete. We will see below that lignified walls are responsible for the strength of wood.

primary and secondary wall

suberin and lignin

Note that cells which are fully lignified or fully suberised are dead because of the waterproofing property of lignin and suberin.

1.3.2 The occurrence of pits

pits

An examination of Figure 1.1 shows the presence of plasmodesmata, the tubular connections between adjacent cells. They will be dealt with in detail later in the chapter. During the process of cell wall formation no secondary wall is laid down immediately over the plasmodesmata and such regions are called pits (Figure 1.4a).

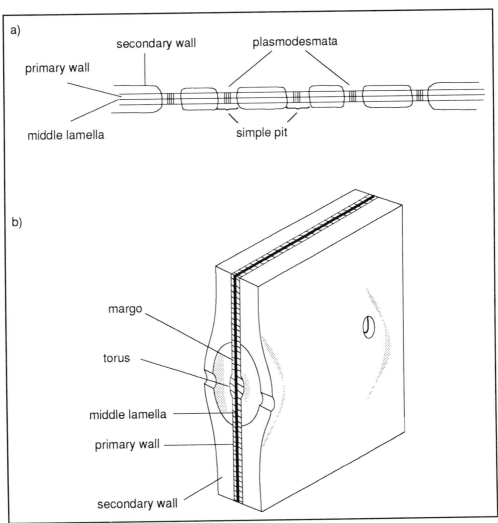

Figure 1.4 Diagram of a cell wall to show a) simple pits and b) a bordered (torus) pit. The plasmalemma has been omitted for clarity.

The gaps in the secondary wall usually occur in pairs, one on each side of the cell wall, forming what are called simple pit pairs. Lignin is not laid down in pits. In some species, particularly gymnosperms, the secondary wall overreaches the plasmodesmata to form a bordered pit (Figure 1.4b). The portion of primary wall running through a pit is called the pit membrane and in bordered pits it is often swollen in the centre, forming the torus. The surrounding pit membrane, the margo, is flexible and if there is a pressure difference across the pit pair, as when an air bubble forms on one side, the torus is forced against the border, effectively sealing it. This mechanism is important in xylem cells, as we shall see in a later chapter. Let us attempt an SAQ before proceeding further.

Match the word or phrase in List 1 with those in List 2. Words or phrases in List 2 may be used more than once.

List 1

a) Middle lamella.

b) Simple pits.

c) Primary wall.

d) Torus pit.

e) Secondary wall.

f) Waterproofs the wall.

List 2

1) Lignin.

2) Prevents lateral spread of air bubbles.

3) Are present in growing cells.

4) Contain both hemicellulose and cellulose.

5) Suberin.

6) Present only in cells which have stopped growing.

7) Is the material which holds cells together.

8) Regions where the secondary wall is absent.

1.4 Plants are composed of three main tissues

plant tissues

Plant organs are all made of cells, of course, but we recognise a level of organisation between cells and organs termed 'tissues'. A tissue consists of a number of cell types which usually cooperate in a single function. An organ can be said to be a discrete entity which contains a number of tissues, an example being a leaf. Plants have three tissue systems: dermal tissue, ground tissue and vascular tissue.

1.4.1 Dermal tissues

epidermis

The term dermal tissue is applied in the main to young plants. In such plants there is a single, clearly defined, single-celled, outer layer, called the epidermis. The epidermis is retained as the outermost layer of the leaf and, since leaves tend to be repeatedly made afresh, even in evergreens, most leaves tend to be less than two or three years old. However, the stems and roots of perennials persist for many years and the original epidermis becomes replaced by the bark, which is, strictly speaking, not a dermal tissue.

trichomes

stomata

The shoot epidermis, ie that of the leaf and the stem, is usually coated with a waxy layer, the cuticle, which reduces water loss. Hairs, called trichomes, are often present and these may consist simply of extensions from single epidermal cells or be multicellular in structure (Figure 1.5a). The epidermis in the root does not have a cuticle and this aids in its function of water and mineral absorption. Root hairs, extensions from epidermal cells, are present close to the growing point of roots (Figure 1.5b). There is one further specialisation of the dermal tissue that we need to mention and these are the guard cells and stomata (Figures 1.5c and d).

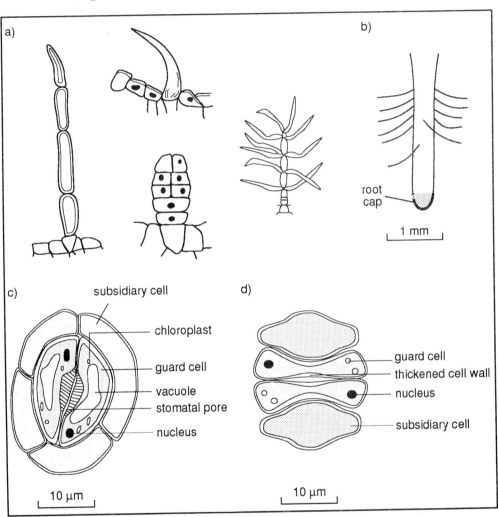

Figure 1.5 Specialisation of dermal tissues: a) trichomes; b) root hairs; c) eliptical guard cells; d) graminaceous guard cells. Gramminaceous guard cells often contain starch grains.

Present in all aerial epidermes are pairs of cells which have a slit-like microscopic pore in the middle lamella of their common wall. The pore is a stoma (plural stomata) and the cells around it are called guard cells. The pore allows gaseous exchange between the outside atmosphere and the interior of the plant. The guard cells are usually smaller than the other epidermal cells and they show structural specialisation in relation to their function. We will examine the details of these in a later chapter.

We can see, then, that the dermis forms the 'skin' of the plant. The interior of the plant consists of the vascular tissue embedded in ground tissue.

1.4.2 Ground tissues

parenchyma

Ground tissues make up the bulk of young plants but occupy a smaller proportion of the plant mass as development proceeds. Three cell types are recognised within ground tissues; parenchyma, collenchyma and sclerenchyma, and of these parenchyma cells are the most abundant. These cells are referred to as ground tissue because they are the packing cells which fill the space between other tissues. They are the least specialised cells but, since they account for more than 80% of all the cells in the plant, they are of considerable importance. Parenchyma cells (Figure 1.6a) are typically 15μm wide by 30-40 μm long, highly vacuolated and usually with thin walls. The cells are living when mature and retain their nucleus. Parenchyma cells in leaves are characterised by the presence of large numbers of chloroplasts and these cells are sometimes referred to as chlorenchyma. Chlorenchyma often contain starch grains.

chlorenchyma

cortex

aerenchyma

The cortex of stems and roots is predominantly composed of parenchyma and they are often used to store starch. Tissues made up of parenchyma are usually well supplied with air spaces. In many aquatic species air spaces between parenchyma cells are very large indeed, often occupying more space than the cells themselves. In such cases the cells are referred to as aerenchyma.

collenchyma

Collenchyma cells (Figure 1.6b) are found in young tissues, are typically longer and narrower than parenchyma cells and have walls which are thickened, with extra cellulose often in the corners of the cell. They are found towards the periphery of herbaceous stems, petioles and in the ribs of leaves. Often groups of collenchyma cells form distinctive ribs or outer edges of stems or petioles as in celery and rhubarb or stem of nettle. Thus, they are considered to have a structural role. Note that the strength of collenchyma tissues rests solely on the strengthening property of cellulose.

sclerenchyma

Sclerenchyma cells also have thick walls but these are impregnated with lignin, which as we saw earlier, considerably strengthen cell walls. Sclerenchyma cells are dead when mature, and are divided into two types; fibres, which are long and thin, and sclereids, which are approximately isodiametric and sometimes branched. Fibres are usually approximately 5-10 μ in diameter but can be as much as 1 mm in length. They are often found in association with vascular tissue, particularly supporting the phloem, where they often occur as multicellular strands. These strands are very prominent and in certain species, such as flax, *Linum usitatissimum* and hemp, *Cannabis sativa* they are extracted and woven into linen or rope. Sclereids may occur in layers, as in the formation of hard seed coats, but they also occur in groups with no obvious function, such as the 'stone' cells of pear fruit tissue, in which they are responsible for the gritty texture.

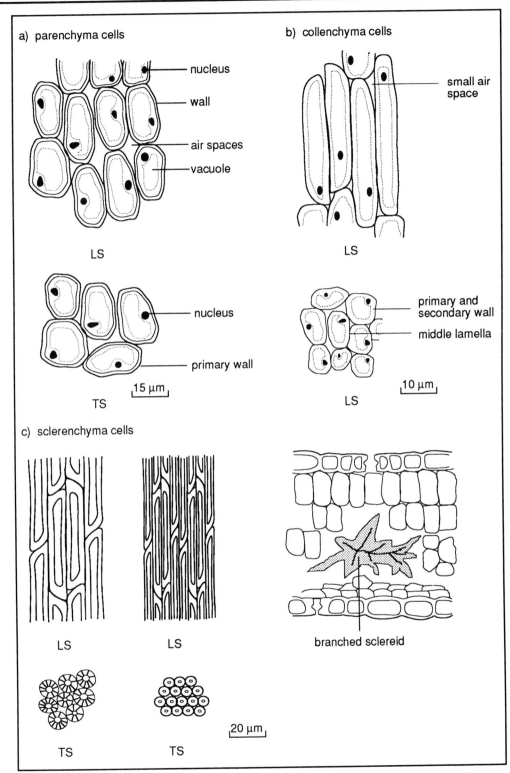

a) parenchyma cells

nucleus

wall

air spaces

vacuole

LS

nucleus

primary wall

15 µm

TS

b) collenchyma cells

small air space

LS

primary and secondary wall

middle lamella

10 µm

LS

c) sclerenchyma cells

LS   LS

branched sclereid

TS   TS

20 µm

Figure 1.6 Diagrammatic representaton of a) parenchyma; b) collenchyma; c) sclerenchyma. Note that both longitudinal (LS) and transverse sections (TS) are shown.

1.4.3 Vascular tissue

*xylem and
phloem*

This consists of two conducting systems. Water and minerals are transported by cells of the xylem, whereas the products of photosynthesis and a number of other organic compounds are transported in the phloem.

Xylem. The conducting cells of the xylem consist of tracheids and vessel members which have in common the fact that their protoplasms die during differentiation. Thus these cells have no cytoplasm or nucleus when mature, and their walls are impregnated with lignin in one of a series of possible forms (Figure 1.7).

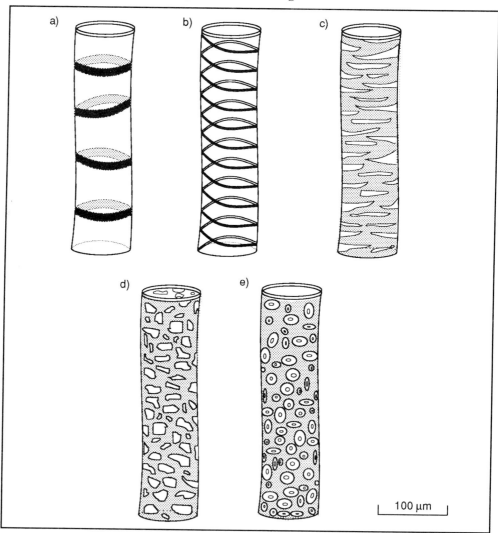

Figure 1.7 Types of cell wall thickening in xylem cells: a) annular; b) helical; c) scalariform; d) reticulate; e) pitted.

Tracheids are very long and thin cells with pointed ends, typically 1-2 mm long by 20µm wide (Figure 1.8). In contrast, vessel members are shorter and wider, being approximately 500 µm in length by 85 µm wide.

Vessel members are aligned in the longitudinal direction and are in contact with each other by their end walls. These are oblique to transverse in orientation and contain numerous elongate perforations forming the perforation plate. The aligned xylem vessel members form the xylem vessel proper, which is often 5 metres long but in tree species, with wide vessels, the xylem cells may extend throughout the entire height of the tree. Usually in trees, the xylem cells are produced in spring.

Tracheids do not have perforation plates and contact from one tracheid to another is achieved through simple pit pairs. The lignified walls of xylem tracheids and vessels are responsible for the strength of wood.

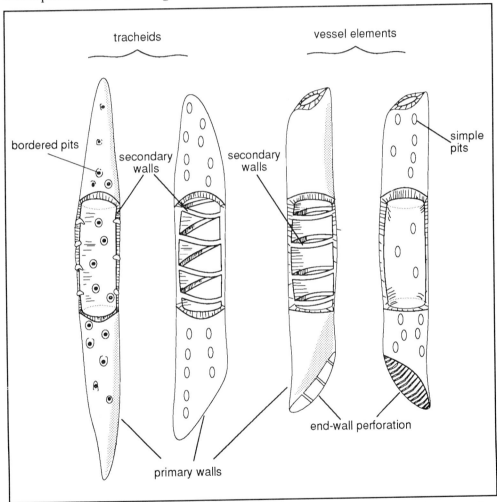

Figure 1.8 Diagrammatic representation of xylem conducting cells. Note that these are not drawn to scale Tracheids are often 1-2mm long and about 20mm wide. The vessel members are typically 0.5mm long and 85 mm wide. Part of the walls have been cut away to show how the walls have been thickened.

phloem cells

Phleom. The cells of the phloem which are specialised for transport are the sieve-tube members which are approximately 200 μm long by 15 μm wide. Their end walls tend to be transverse in orientation and to be perforated, not by elongate perforations as in xylem vessels, but by circular perforations Figure 1.9 shows a non-circular-perforated sieve plate.

Sieve cells are aligned longitudinally and these form very long sieve tubes, with access from one cell to the next being gained through the sieve plate. Although there are some similarities between phloem sieve tubes and xylem vessels, there are some very important differences. The major difference is that sieve tube cells are living cells. However, during their differentiation, their nuclei degenerate and their metabolism is considered to be directed by the nucleus of a companion cell (Figure 1.9), a cell type normally found in close association with a sieve tube cell.

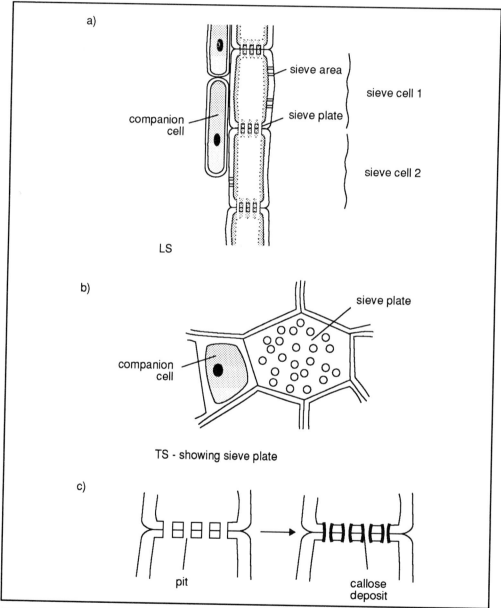

Figure 1.9 Diagrammatic representation of sieve and companion cells. In a) we have shown a longitudinal section of two sieve cells together with a companion cell. Note that sieve plates occur at each end of the sieve cells. There are also sieve areas in the longitudinal walls of the sieve tubes which allows for lateral movement of water. In b) we illustrate a transverse section showing a sieve plate. In c) we show a longitudinal section of the sieve plate. As the sieve plate ages callose is deposited in the pits and eventually the pits may become closed.

In addition to the specific conducting cells described already, xylem and phloem tissues also contain parenchyma cells and sclerenchyma fibres.

Let us try some SAQs before proceeding to the final section of this chapter.

SAQ 1.3

Match the word or phrase in List 1 with those in List 2. The latter may be used more than once.

List 1		List 2	
a)	dermal tissue	1)	parenchyma
b)	vascular tissue	2)	xylem tracheids
c)	ground tissue	3)	stoma
		4)	epidermis
		5)	sieve tube cell
		6)	collenchyma
		7)	chlorenchyma

SAQ 1.4

Name two commercially-important plant products which depend directly on lignified tissues. Give the name of the cell types involved and decide whether they show higher or lower rates of respiration, when mature, than other cells. Explain the last part of your answer.

SAQ 1.5

Match the word or phrase in List 1 with those in List 2, which may be used more than once.

List 1		List 2	
a)	is found in xylem tissues;	1)	collenchyma;
b)	is found in phloem tissues;	2)	tracheid;
c)	is the least specialised plant cell;	3)	sieve tube cell;
d)	has walls thickened in the corners of the cells;	4)	vessel element;
e)	contains lignified walls;	5)	parenchyma;
f)	contains suberin;	6)	sclerenchyma fibres;
		7)	cork cells.

1.5 Plasmodesmata and the symplasm/apoplasm system

plasmodesmata

We referred earlier to plasmodesmata, the tubular connections which link adjacent plant cells. Figure 1.10 is a schematic representation of four cells showing how the cytoplasms of the cells of a tissue is linked into a three dimensional network by the plasmodesmata. Also note the air spaces.

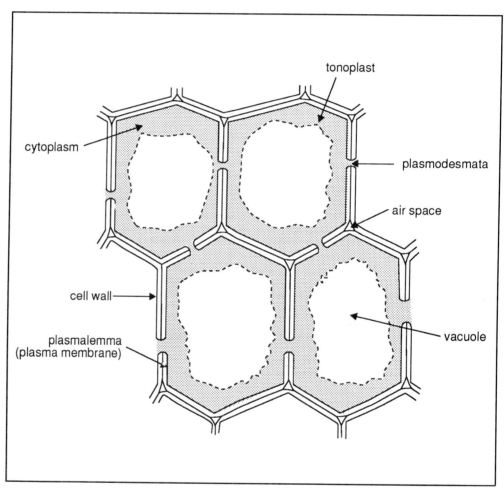

Figure 1.10 Schematic representation of the symplasm/apoplasm system. Only one plasmodesma is shown on each cell wall, the actual number varying from 10-50. The symplasm is stippled, the apoplasm is clear.

symplasm

apoplasm

The vacuole excepted, the symplasm refers to everything bounded by the plasmalemma. The apoplasm refers to everything outside the plasmalemma except the air spaces entrapped between cell walls. Generally speaking, it is considered that low molecular mass compounds (those less than 800 Daltons) can diffuse freely within the apoplasm or within the symplasm and that these two routes constitute important transport routes within plant tissues. This point is illustrated in Figure 1.11.

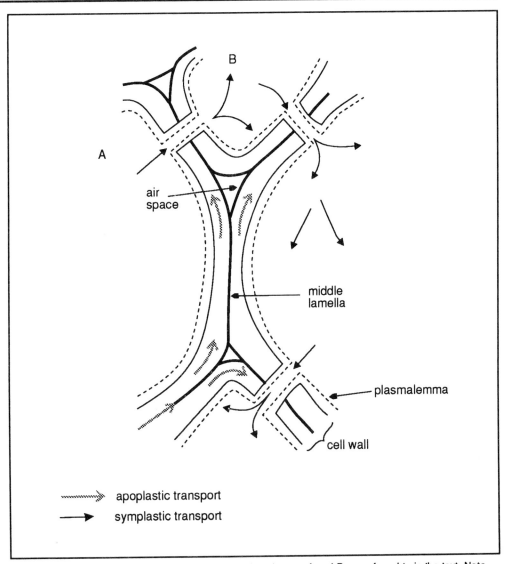

Figure 1.11 Apoplastic and symplastic transport in plant tissues. A and B are referred to in the text. Note that movement is not unidirectional but can occur in both directions. Light arrows indicate apoplastic transport, dark arrows indicate symplastic transport.

If a low molecular mass solute is present in the cytoplasm of cell A (Figure 1.11), it will be able to move into cell B by diffusing through the plasmodesmata. This process does not require energy. However, if there were no plasmodesmata connecting the two cells the solute would have to pass through the plasmalemma and cell walls of cell A and then the walls and plasmalema of cell of B. Assuming translocators were present which could assist these processes, such a movement would consume energy. Thus, plasmodesmata allow 'free' solute transport, ie transport which does not consume energy.

The apoplast constitutes a separate system through which low molecular mass compounds can diffuse and we will see examples in later chapters of the importance of this. However, it is not difficult to understand that the totally free diffusion of solutes through the apoplasm might present problems. Think of toxic metals present in the soil; if these could diffuse freely through the apoplasm all over the plant this might be very **apoplastic** detrimental for the delicate tissues of the shoot. This is prevented by a root apoplastic **barrier** barrier, the Casparian strip, which will be described when we discuss mineral nutrition and transport in Chapter 4.

∏ See if you can think of a way in which a small molecular mass compound might be prevented from diffusing from one cell to another in the symplasm.

There are various ways in which this might be achieved. Binding solute to a protein or other large molecular mass molecule could make it too big to pass through a plasmodesma. Binding to the endoplasmic reticulum (ER) , or any other organelle would have the same effect. A commonly employed method to achieve this in plant roots is the sequestering of the compound in the vacuole, thus taking it out of the diffusion stream.

| **SAQ 1.6** | Can proteins diffuse freely through the apoplasm? |

The foregoing discussion may have given the impression that plasmodesmata are simply open tubes but it is now clear that this is not the case.

desmotubule Electron micrographs sometimes show open tubes but in most cases there is electron-dense material inside the plasmodesma itself. Several clear pictures have been obtained showing continuity between the endoplasmic reticulum (ER) and a membranous tube, the desmotubule, which runs through the plasmodesma. These observations have been used to produce the scheme shown in Figure 1.12b.

The desmotubule does not always appear to be present, and most micrographs of plasmodesmata do not allow the internal structure to be discerned at all. We would assume, however, that all plasmodesmata have the structure shown in Figure 1.12b and that unclear micrographs are due to non-central sectioning and, or variations in tissue fixation. It is therefore implied that there are two channels running from one cell to the next. One of these is outside the desmotubule and links the cytosol of adjacent cells. The other is inside the demotubule and links the lumen of the ER of one cell with that of the adjacent cell. Although overall dimensions of plasmodesmata suggest that they are approximately 50 nm in diameter, experiments following the transport of fluorescent dyes suggest that the functional diameter is only about 7 nm. Thus the exclusion limit **pore diameter** appears to be a molecular mass of approximately 10,000 Daltons. However, further **may be variable** recent evidence suggests that the diameter of the pore can vary quite significantly. It is known that viruses move from one cell to another through plasmodesmata and they appear to induce an increase in the diameter of plasmodesmata.

On the other hand, the rate of diffusion of permeant fluorescent dyes varies with the metabolic state of the tissue. It is considered that some of the observed changes in plasmodesmal transport involve changes in the diameter at the neck region and it has been proposed that a protein sphincter controls this. The idea that plasmodesmata can be opened or closed is very exciting for it has a bearing on our understanding of the control of short distance transport, a topic that will be covered in later chapters.

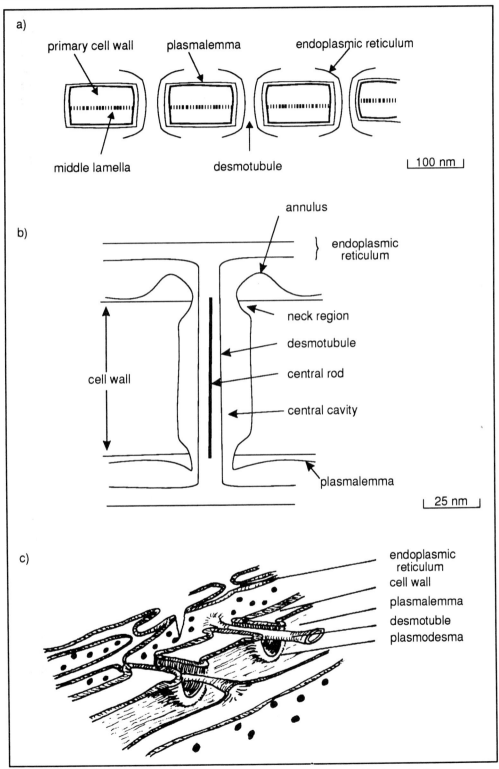

Figure 1.12 Diagrams of plasmodesmata: a) wall showing plasmodesmata, b) schematic diagram of a single plasmodesma, c) three dimensional representation of the symplastic links between cells.

Summary and objectives

The chapter began with an overview of plant cell ultrastructure. General plant morphology was then described and the position and roles of root and shoot meristems explained. A simple account was given of cell wall structure paying particular attention to the middle lamella, primary wall and secondary wall layers and the formation of pits, as a basis for understanding the differences between cell types. Plants are organised into cells, tissues and organs and three main tissues are recognized, dermal, ground and vascular tissues. The major cell types are parenchyma, collenchyma, sclerenchyma, xylem vessels and tracheids and phloem sieve cells and their major structural features were described. Plasmodesmata were described and an outline given of the structure of the symplasm/apoplasm system. The chapter finished with a description of current views of plasmodesmatal structure and of possible ways in which solute movement through them might be controlled.

Now that you have completed this chapter you should be able to:

- describe the function and location of apical shoot and root meristems, axillary buds and lateral roots, and give a general account of plant morphology;

- explain the term apical dominance;

- describe the three layers which may be present in a cell wall and explain what pits are;

- show an understanding of the fundamental organisation of plant cells into tissues and organs;

- demonstrate an understanding of the nature of dermal tissues, ground tissues and vascular tissues by describing the structural features of cells found in them;

- describe what is meant by the terms apoplasm and symplasm;

- describe current views of the structure and function of plasmodesmata.

Photosynthesis

Photosynthesis

2.1 Introduction

definition

In the previous chapter, we examined a number of aspects of plant structure and function to provide a basis for understanding how plants work. The story of plant physiology proper begins here. Photosynthesis, the process where by plants, algae and some bacteria are able to convert solar energy into chemical energy, is arguably the most important biological process on planet Earth. Although a number of organisms are capable of photosynthesis, in terms of the biomass produced, green plant photosynthesis is by far the most important. We will be concerned almost exclusively with terrestrial plants and we will be following a theme. Land plants evolved from aquatic predecessors and various changes were necessary for plants to become successful in a terrestrial environment. Bearing in mind that the primary aim of plants is to survive and reproduce, our theme will be to identify the physiological problems posed by the terrestrial environment and to examine the ways in which they have been overcome. Obviously, the problems have been solved, for if not, plants would not exist today and nor, for that matter, would we.

2.2 Photosynthesis occurs in chloroplasts

chloroplast

thylakoids

All of the reactions of photosynthesis take place in the organelle known as the chloroplast. Except for the nucleus, the chloroplast is by far the largest plant cell organelle. It is ellipsoidal in shape and is approximately 7.5 μm long and 3 μm wide. The most striking feature of its structure is the system of internal membranes termed thylakoids. These form sheets which run parallel with the long axis of the chloroplast and, in places, are stacked up on top of one another to form what are called grana. A granum consists of a number of thylakoids, each with a distinct intra-thylakoid space or lumen (Figure 2.1).

appressed
thlakoids

In places, the thylakoids extend laterally and join up with those in an adjacent granum. These inter-granal thylakoids result in an intra-thylakoid space continuous with several thylakoids. Most thylakoids in a granum, however, are not so linked; their intra-thylakoids spaces are isolated. Thylakoids in grana are called appressed thylakoids. The space around the thylakoids in the chloroplast is filled with a fluid called the stroma and the structure is completed by the presence of two limiting chloroplast envelope membranes with a small inter-membrane space (Figure 2.1).

envelope
membranes
and stroma

Thus, we can see that the chloroplast contains three membrane-bound compartments; the envelope intermembrane space, the stroma and the thylakoids. We will see later that the outer envelope membrane of the chloroplast is generally permeable to solutes of small molecular mass but the inner envelope membrane and thylakoid membrane are not. This control over permeability is essential for chloroplast function.

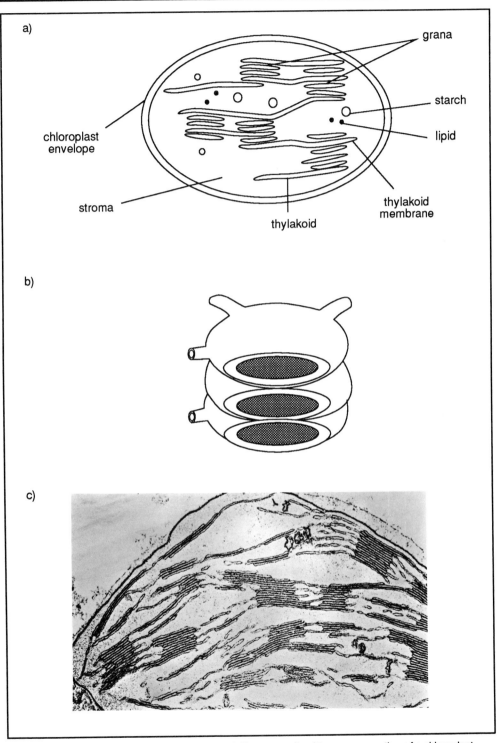

Figure 2.1 Membrane systems of a chloroplast. a) Shows a stylised transverse section of a chloroplast. Note that each chloroplast is surrounded by a double membrane. Within the chloroplasts, thylakoids, stacked into grana, are comprised of membranes in the form of vesicles. The thylakoid membrane may extend from one granum to another. A stylised representation of a granum with its interconnecting thylakoid membranes is shown in b). c) Transmission electron micrograph of a chloroplast.

2.3 Photosynthesis simply stated

The overall reaction for photosynthesis can be written as follows:

$$CO_2 + H_2O \xrightarrow{\text{light}} (CH_2O) + O_2$$

(CH_2O) is an abbreviation for carbohydrate.

Remember that light behaves as if it consists of discrete packets of energy called quanta. About 9 to 10 quanta are needed to drive the reaction depicted which has a Gibbs free energy change of +467 kJ mol^{-1}. Note that the Gibbs free energy change is sometimes referred to as Gibbs function change.

$\prod$ The energy in a quantum of light is related to its wave length by $E = hc/\lambda$ where E = energy, h = Planck's constant, c the speed of light and λ = wavelength of the light. Given that $h = 6.62 \times 10^{-34}$ Js, c is $3 \times 10^8 ms^{-1} (3 \times 10^{17}$ nm s$^{-1})$ and 1 mol of quanta contains 6.023×10^{23} quanta, calculate the energy of a mol of red light, wavelength 680nm. Assuming that 10 quanta of red light are used for each molecule of O_2 released, calculate the efficiency of the conversion. (Try this before reading our solution).

Energy of 1 quantum of red light:

$$\frac{6.62 \times 10^{-34} \text{ Js} \times 3 \times 10^{17} \text{ nm s}^{-1}}{680 \text{ nm}}$$

$$= 0.0292 \times 10^{-17} J.$$

Since there are 6.023×10^{23} quanta in a mol of quanta.

Energy of 1 mol of quanta $= 0.0292 \times 10^{-17} \times 6.023 \times 10^{23} = 176$ kJ mol^{-1}.

10 mol of quanta are required to release 1 mol of O_2 .

Therefore 10 mol of quanta $= 1760$ kJ mol^{-1}.

But we said that the change in Gibbs function for the conversion of $CO_2 + H_2O$ to fixed carbon and O_2 was + 467 kJ mol^{-1}.

Therefore efficiency $= \dfrac{467 \times 100\%}{1760} = 26.5\%.$

In 1931 Van Niel drew an analogy between photosynthetic bacteria and green plants by suggesting that the equation for both could be written as:

$$CO_2 + 2 H_2A \longrightarrow (CH_2O) + 2A + H_2O$$
$$\text{reduced donor} \qquad \text{oxidised donor}$$

If this was the case, it would suggest that the O_2 released by plants originated in the molecule of water so the equation should, theoretically, be written as:

$$CO_2 + 2H_2O \longrightarrow (CH_2O) + O_2 + H_2O$$

This point was confirmed only when radioisotopes became available. Later work by Blackman on the effects of temperature and irradiance showed that part of the overall process was purely chemical in nature. This laid the foundation for the division of photosynthesis into the so called light and dark reactions. Subsequent work by Hill and by Arnon suggested that the light reaction involved the splitting of water to form ATP and reduced nicotinamide adenine dinucleotide phosphate. $NADP^+$ is the acceptor compound:

light reactions

$$H_2O + NADP^+ + ADP + Pi \xrightarrow{\text{light}} O_2 + ATP + NADPH$$

dark reactions

It was logical to suggest that in the dark reaction CO_2 was reduced using this energy and reducing power:

$$CO_2 + ATP + NADPH \longrightarrow (CH_2O) + ADP + Pi + NADP^+$$

and these suggestions were quickly confirmed. We have represented this relationship between light and dark reactions in Figure 2.2.

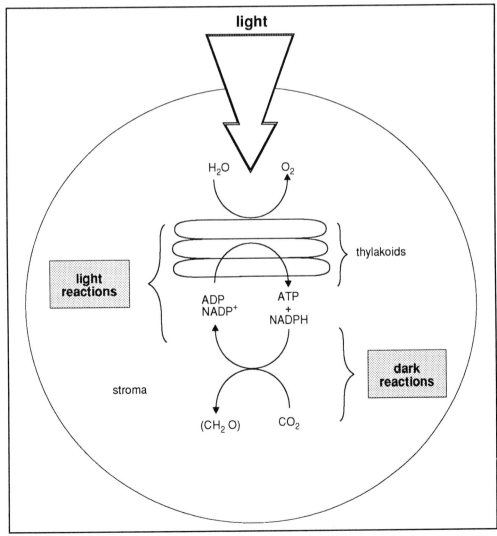

Figure 2.2 In the chloroplast ATP and NADPH are formed in the light reactions and CO_2 is reduced in the dark reactions.

A word of caution needs to be added here. It is usual at this stage in the description of photosynthesis to state that whereas the light reaction is light requiring the dark reaction is not dark requiring. Thus the suggestion is made that light-dependent and light-independent would be better terms. However, we know that several of the enzymes which function in the reduction of CO_2 are indirectly activated by light. Thus, the dark reactions are not light-independent. However, only in the light reactions is the energy of light converted into a chemical form, and as long as we remember this the use of the terms light and dark reactions should not pose a problem.

Let us check your understanding of the division into the light and dark reactions with an SAQ.

SAQ 2.1

Fill in the gaps in the boxes in the scheme below which describes photosynthesis.

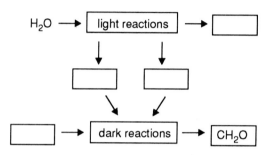

2.4 Two photosystems are involved in the light reactions

In this and the next section we will examine the light reactions to see how O_2, NADPH and ATP are formed. We will deal with this in two parts; in the first, we will discuss the formation of O_2 and NADPH, and in the second, the formation of ATP.

2.4.1 O_2 and NADPH production

photosystems

A photosystem is a large collection of pigment-protein complexes, often of molecular masses approaching 500,000 Daltons, which function to boost electrons from low to high energy at a special pigment-protein termed a reaction centre. Two photosystems interact in photosynthesis, as shown in Figure 2.3.

reaction centres

PSI and PSII are photosystems whose reaction centres utilise light of wavelength 700 and 680 nm respectively. (Note PSI = Photosystem I and PSII = Photosystem II). These reaction centres are termed P700 and P680. Phaeophytin (Phae) is a chlorophyll molecule lacking a magnesium atom. Q_A, and Q_B are quinones bound to protein which forms a part of the PSII complex. Q_A is firmly bound but Q_B is bound when oxidised and mobile when reduced. In its mobile form Q_B is referred to as plastoquinone (PQ). CYT is a complex containing cytochrome b and f molecules. PC is a protein known as plastocyanin. Fd1, 2 and 3 are iron-sulphur centres called ferredoxins. Fdl is part of the P700 reaction centre, and the other two reside on an associated protein. SFd is a soluble ferredoxin.

When PSII absorbs light an electron is transferred from P680 to Phae and thence to Q_A, Q_B and PQ. PQ transfers its electrons to CYT after which they are transferred to PC. Meanwhile PSI has absorbed light and transferred an electron via Fd1, 2 ,3 and SFd to $NADP^+$. P700 and P680 are each left with a positive charge. P680 is brought back to neutrality by accepting an electron from the splitting of water. P700 regains neutrality by accepting an electron from PC. The products which can be identified so far are O_2, from the splitting of water, and NADPH. It can be seen that the two photosystems work in series. They are also inter-dependant. Thus, PSI releases an electron to Fd1, but it cannot release another one until it has received the electron from PC. Similarly, PSII activation will cease if Phae is not oxidised by passing its electron through the electron transport chain to PSI.

water splitting

NADP $^+$
reduction

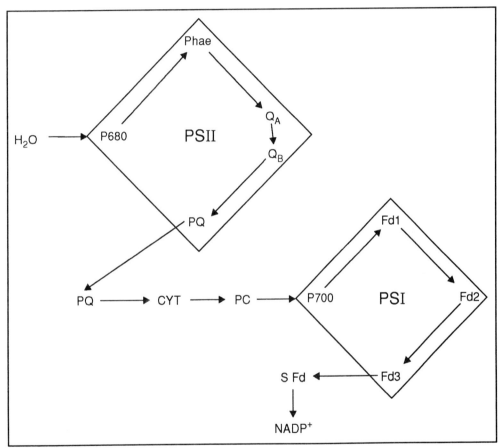

Figure 2.3 Pathway of electrons in the light reactions to show the interaction of two photosystems. See text for details.

Ⅱ We have given you rather a lot of chemicals and their abbreviations to remember. Perhaps it would be helpful to re-read the last section and to make a summary table of these abbreviations for example: PSI = photosystem I.

Ⅱ If we refer to PSII as water-plastoquinone oxidoreductase, what could we call a) PSI and b) the electron transport chain?

These would be a) plastocyanin-NADP$^+$oxidoreductase and b) plastoquinone-plastocyanin oxidoreductase.

SAQ 2.2

What prediction would you make concerning the oxidation state of the cytochrome b/f complex shown in Figure 2.3 if the system was irradiated with light of:

1) wavelength 680nm alone?

2) wavelength 700nm alone?

Explain your answers.

SAQ 2.2 is based loosely on one of the pieces of evidence which led to the proposal of the scheme shown in Figure 2.3. Two other pieces of evidence will now be considered briefly.

2.4.2 The red drop and the enhancement effect

quantum yield

The quantum yield of photosynthesis is the reciprocal of the number of quanta needed to reduce one molecule of CO_2 to carbohydrate. It is based only on the quanta which are absorbed, is fairly constant for light between 400 and 680nm but falls off dramatically above 680nm. Light with a wavelength of greater than 680nm is much less efficient at photosynthesis than light of lower wavelength. This phenomenon is called the red drop and refers to the drop in efficiency in the red end of the spectrum (Figure 2.4a).

red drop

enhancement

Experiments carried out by Emerson in the 1950s showed that the efficiency of photosynthesis could be extended at the red end of the spectrum by the addition of supplementary light of wavelength 650nm. Further, the rate of photosynthesis supported by red light (600-680nm) or by far-red light (680-750nm) was very strikingly stimulated when light of both are applied together. The action of the two together was more than additive (Figure 2.4b). This phenomenon is referred to as the enhancement effect.

The red drop and the enhancement effect were very puzzling and the proposal of the scheme in Figure 2.3, with the two photosystems working in series, was the most logical explanation.

2.4.3 ATP production requires a proton gradient

thylakoid protein complexes

In order to explain how ATP is formed we need to examine the protein complexes of the thylakoid in more detail and to see how they are arranged on the thylakoid membranes. The thylakoid contains five major protein complexes:

*   photosystem I;

*   photosystem II;

*   cytochrome b/f complex;

*   oxygen evolving complex (OEC);

*   ATP synthase.

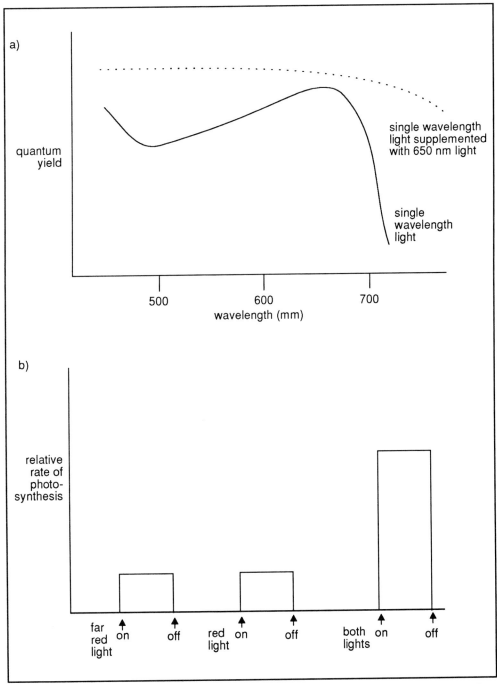

Figure 2.4 a) Reduction in efficiency of photosynthesis at the red end of the spectrum(——) and effect of supplementary 650nm (····) light. b) Effect of the simultaneous application of red and far-red light on photosynthesis compared with the two colours applied separately.

As shown in Figure 2.5, photosystems I and II and the cytochrome complex span the thylakoid membrane. The OEC is located on the lumen face of the membrane, in association with PSII. ATP synthase consists of two parts; CF_0, a hydrophobic complex which binds to CF_1 on the stromal face.

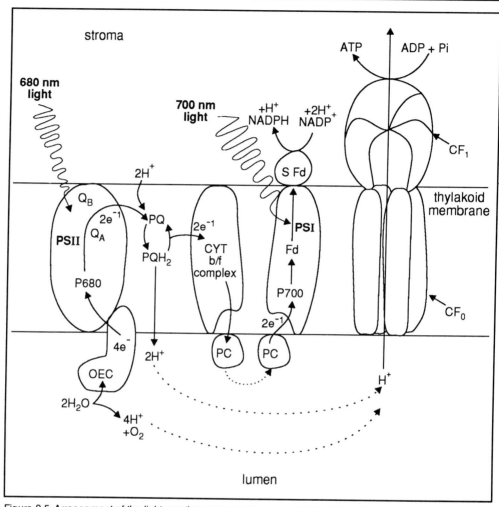

Figure 2.5 Arrangement of the light reaction components on and in the thylakoid membrane (see text for details).

PSII spans the thylakoid membrane

NADPH production in the stroma

oxygen evolving complex in the lumen

The membrane represented in Figure 2.5 consists of the normal phospholipid bilayer with various components embedded in it or bound to its surface. Note that the stroma is to the top and the thylakoid lumen to the bottom of the diagram. The PSII complex spans the membrane with the Q_A- protein located on the stroma side of the membrane. When the Q_B protein has accepted two electrons from PSII it absorbs $2H^+$ from the stroma and PQH_2 is released into the membrane. Note that PQ is a hydrogen as well as an electron carrier and needs both in order to be reduced. Each photoact releases one electron from PSII. Therefore, two photoacts are required to produce the two electrons needed to reduce PQ. PQH_2 reduces the cyt b/f complex but because this complex accepts only electrons, the protons are released again. However, the geometry of the cyt b/f complex is such that the protons are released into the lumen.

The electrons are passed from cyt b/f to PC, which is on the lumen side of the membrane, from which they are transferred to PSI. Each photoact by PSI also releases one electron; thus two are needed to produce the two reduced SFd molecules which are used to produce NADPH + H^+. Fd 2 and 3 are bound to a protein which abuts onto the stroma. SFd is a soluble component in the stroma as is the enzyme which catalyses the reduction of NADP$^+$. Thus, NADPH + H$^+$ is produced in the stroma. Figure 2.5 differs from Figure 2.3 in two important ways. Firstly, it shows the location in the thylakoid of the activities described so far and secondly it shows the release of protons into the thylakoid lumen. It is these which are used in the production of ATP. Protons are released not only in the oxidation of PQH$_2$ by cyt b/f but also by the splitting of water by the OEC. The OEC is situated on the thylakoid lumen . It utilises the strong oxidising power of P680$^+$ to split water and results in the release of protons into the lumen. The thylakoid membranes are permeable to O$_2$.

proton gradient

coupling factor

The light-driven proton accumulation lowers the pH of the thylakoid space very considerably, often to a figure as low as pH 4 which is approximately 4 pH units lower than in the stroma. It is this transmembrane protonmotive force which drives photophosphorylation. The phosphorylation of ADP is catalysed by a membrane-bound enzyme complex which couples ATP formation to proton transport. The complex called coupling factor is a large (323kD) protein complex. It consists of a hydrophobic part (CF$_0$) which spans the thylakoid membrane and has bound to it a hydrophobic protein (CF$_1$), which is a 10nm particle protruding into the stroma. CF$_0$ transfers protons from the thylakoid space to CF$_1$ and this action provides energy which drives the conversion of ADP to ATP. The exact mechanism of this is not known at present. The CF$_0$/CF$_1$ complex acts as an ATPase because its activity is reversible. In the presence of light, the proton gradient is produced and the complex functions to generate ATP. Approximately 3 protons are needed to generate one ATP. Four photoacts by each photosystem produce 1 molecule of O$_2$, 2NADPH and 8 protons. At least 2 ATP are produced from these protons but there is disagreement as to whether or not 3ATP could be formed. Thus, ATP and NADPH are produced in at least equal amounts but there does not appear to be enough protons to ensure the production of 3ATP for every 2 NADPH. The significance of this point will be explained later.

SAQ 2.3

Chloroplasts were carefully isolated from spinach leaves and lysed so as to release the thylakoids and several experiments were carried out.

1) The isolated thylakoids were incubated for several minutes in light in a medium at pH 4.0 and then transferred to a medium at pH 8.0. ADP and Pi were added shortly after and analysis showed that ATP had been formed.

2) The experiment described in 1) was repeated but 2,4-dinitrophenol was added on transfer to the pH 8.0 medium. Dinitrophenol is a membrane-permeable hydrogen carrier. On addition of ADP and Pi a much reduced amount of ATP was formed.

3) The experiment described in 1) was repeated but the thylakoids were inadvertently treated very roughly. This did not break open the thylakoids but it was noticed that numerous 10nm particles were floating freely in the suspension. After pH 4.0 and pH 8.0 treatment as before, addition of ADP and Pi resulted in very little ATP formation.

Suggest explanations for these observations.

2.4.4 Chlorophyll molecules are arranged into light-harvesting complexes

Figures 2.3 and 2.5 indicate how light of wavelengths 680 and 700nm may be utilised in photosynthesis. If these were the only two wavelengths utilised, photosynthetic productivity would be very low indeed. Plants can obviously use extensive zones of the visible light spectrum, as is shown by the action spectrum of photosynthesis (Figure 2.6) and this is clearly related to the absorption of light by the plant pigments.

action spectrum

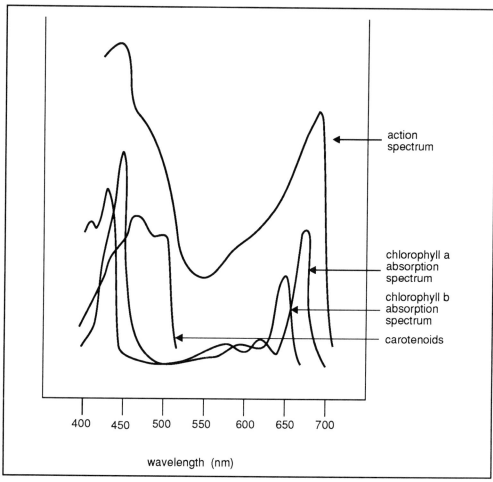

Figure 2.6 Action spectrum of photosynthesis and absorption spectrum of plant pigments.

pigments

light-harvesting complexes

Photosynthetic pigments are all conjugated (alternating single and double bonds) coloured molecules and are bound to thylakoid membranes. They can be divided into three functional groups; the reaction centres of PSI and PSII and the light-harvesting complexes (LHC). The reaction centres of PSI and PSII both contain chlorophyll a molecules complexed with a protein.

antenna pigments

Free chlorophyll a in an organic solvent shows absorption maxima in the blue and red end of the spectrum, at approx 420 and 660nm. Formation of a complex with protein modifies this; the chlorophyll a-protein complex of PSI absorbs maximally at 700nm, whereas the absorption maximum of PSII is at 680nm. The LHC contain chlorophyll a, chlorophyll b and carotenoids, complexed with several different proteins which, as before, modify the absorption properties of the pigments. The presence of LHC around the reaction centres of PSI and PSII broadens the range of light which can be used to drive photosynthesis and for this reason they are often referred to as the antenna pigments or light harvesting complexes.

Let us now examine how the antenna pigments work.

As you can see from Figure 2.6, chlorophyll molecules can absorb light in the blue and the red end of the spectrum but we know that there is considerably more energy in each quantum of blue light than of red. Red light raises an electron from chlorophyll to the first excited state but blue light raises it to the second or even higher states. However, a blue light-excited chlorophyll molecule will quickly revert to its first excited state (which is its lowest excited state) losing the extra energy as heat. Thus, the energy available for photosynthesis from blue and red light is essentially the same (Figure 2.7a).

excitation energy

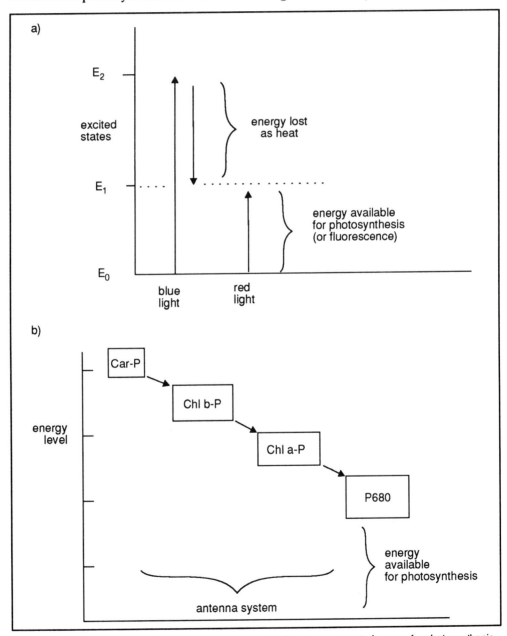

Figure 2.7 a) Blue light and red light provide approximately the same amount of energy for photosynthesis. b) Energy is transferred from pigment to pigment within the antenna system. Note that the pigments are all complexed with proteins, as indicated by the P of Car-P, etc. Car= carotenoid, Chl = chlorophyll.

resonance transfer

Energy can move between pigment complexes by the process of resonance transfer but certain conditions have to be met. On the one hand, transfer requires that the energy absorption bands of the two pigments overlap. This is because the energy available must match possible energy states of the receiving molecule. However, due to a small loss of energy at each transfer the available energy becomes progressively reduced until it reaches the reaction centre (Figure 2.7b). Resonance, therefore, occurs down an energy gradient and is a unidirectional process.

The other consideration is orientation and distance between pigment molecules. The closer the molecules are together the faster and more efficient the transfer, and molecules must be less than 10nm apart for transfer to occur at all. Transfer of energy must occur within 1 nanosecond or else the energy is lost as heat or as light. Note that in both photosystems, the reaction centre chlorophyll is always at the longest wavelength and, therefore, the lowest energy. Energy which reaches PSI or PSII is successfully transferred if the chlorophyll a is not already activated or positively charged. Under conditions of low utilisation of NADPH and ATP, for example, the rate of turnover of PSI and PSII would be low and energy could be available which cannot be used immediately. Much of this energy would be lost as heat or light, the latter being fluorescence.

2.4.5 How is the LHC organised?

Information on this question is available so far only by indirect methods. Experiments using short flashes of light have been used to gain information about the relation between O_2 production and the number of chlorophyll molecules present. Regardless of the amount of light per flash the maximum amount of O_2 produced is 1 molecule of O_2 per 2500 chlorophyll molecules. We have learnt that on theoretical grounds 8 quanta of light are needed to produce 1 molecule of O_2 (four quanta each by PSI and PSII, see Section 2.4.3). This would suggest that each LHC contains approximately 300 chlorophyll molecules, some of which would be chlorophyll a and others chlorophyll b. Carotenoids are also present. Examination of isolated LHC complexes suggest the occurrence of a small number of different chlorophyll-protein complexes. The suggested numbers vary but it is never more than eight. If we take this figure as reasonable, it would suggest that there are possibly 35 copies of each of these in each LHC. Remember that each of the eight different chlorophyll-protein complexes would absorb light at a different wavelength and, in a practical sense, transfer energy one to another only down an energy gradient. For all the chlorophyll molecules to function within a single LHC would require a high level of geometric organisation but we have no information about this at present.

LHC organisation

SAQ 2.4

Answer the following questions, giving reasons for your answers.

1) Blue light contains almost twice as much energy per quantum as red light. Why cannot blue light drive two photoacts in the light reaction?

2) If light of wavelength 660nm and 690nm was shone onto a leaf, to which photosystem would the energy from each be transferred?

3) If light of wavelength 660nm was shone onto a leaf in which NADPH was present in excess what would happen to the absorbed light energy?

2.4.6 Phosphorylation can also occur as a cycle

cyclic and
non-cyclic
photophosphory
lation

Figure 2.5 shows how NADPH, O_2 and ATP can be formed in the light reaction. There is evidence which suggests that it is also possible for ATP alone to be formed and this is a process involving only PSI. In this case, PSI absorbs a quantum of light and SFd becomes reduced, but it does not reduce $NADP^+$. Rather, it migrates such that it can donate electrons to the transfer components linking the two photosystems, the electron being returned to PSI, whence it came. There is some disagreement at the moment as to exactly where in the electron transfer chain the SFd donates electrons. It is considered by some that SFd reduces PQ involving stromal protons thus accounting for a contribution to the proton gradient and, thus, ATP production. Others consider that SFd donates electrons and reduces the cyt b in the cyt b/f complex and this in turn reduces PQ. When PQH_2 is oxidised by cyt f, protons are released into the lumen. The exact mechanism remains to be resolved, but it does not alter the fact that ATP alone can be produced by a light driven process. Because the electron originates at PSI and returns there, and because ADP becomes phosphorylated, the process is called cyclic photophosphorylation. In contrast, the scheme involving both PSI and PSII is referred to as non-cyclic photophosphorylation.

2.5 The dark reactions

As we noted earlier, the dark reactions themselves do not involve reactions in which light energy is converted to chemical energy. However, they do require the products of the light reactions namely NADPH and ATP and so only work in the day time!

2.5.1 Some dark reactions occur as a cycle

carbon fixation

We saw earlier that in the dark reaction, CO_2 was reduced to carbohydrate using NADPH and ATP produced in the light reactions. For many years the way in which this was achieved was a puzzle because scientists were unable to find the compound which initially reacted with CO_2, ie the first product of carbon incorporation. Radiotracer experiments with $^{14}CO_2$ showed that the first stable compound was 3-phosphoglycerate (PGA), a 3-carbon compound, suggesting that CO_2 first reacted with a 2-carbon compound. However, no C_2 molecule could be found to fit in with PGA synthesis. The puzzle was solved with the discovery that the acceptor molecule was the 5-carbon compound, ribulose bisphosphate, and that the scheme operated as a cycle. The cycle is called the photosynthetic carbon reduction cycle (PCR) or the Calvin cycle, after the scientist who headed the team which elucidated it. The cycle as a whole contains a large number of enzyme-catalysed reactions and we are now in a position to describe each of these in detail. However, from the point of view of understanding the role of NADPH and ATP a simplified version of the cycle will suffice. This is shown in Figure 2.8.

PCR (Calvin)
cycle

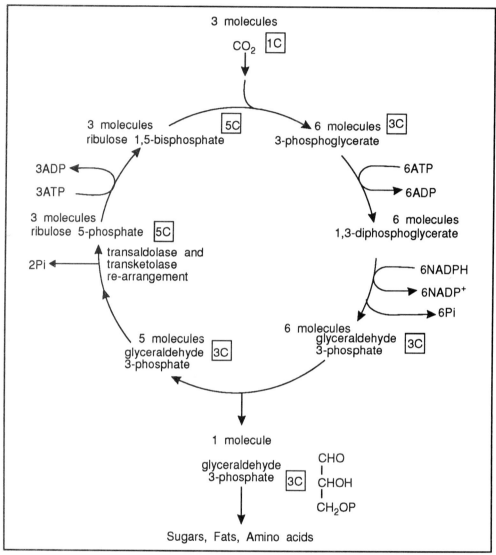

Figure 2.8 A simplified version of the Calvin cycle showing the positions of input of NADPH and ATP. Note that the figures in boxes refer to the number of carbon atoms in the molecules described in the cycle.

Examine Figure 2.8 and write an overall equation to describe it. This should describe the inputs and outputs but not include any input that is reformed in the process as a whole.Use a sheet of paper and begin with the reaction at the top which involves 3 molecules of CO_2 and 3 molecules of ribulose 1,5-bisphosphate. Then work on round the cycle. When you have completed this for all the reactions, add them all together. We have begun this process for you by writing down the first two reactions:

$$3CO_2 + 3 \text{ ribulose 1,5-bisphosphate} \longrightarrow 6 \text{ 3-phosphoglycerate}$$

$$6 \text{ 3-phosphoglycerate} + 6ATP \longrightarrow 6 \text{ 1,3-diphosphoglycerate} + 6ADP$$

The overall reaction is:

$$3CO_2 + 9ATP + 6NADPH \longrightarrow \text{glyceraldehyde 3-phosphate} + 8Pi + 6NADP^+$$

4 phases of the PCR cycle

This shows that 1.5 ATP are required for each NADPH and that 8 rather than 9 Pi are released, because the product is itself phosphorylated.

The scheme shown in Figure 2.8 can be divided into four stages:

- carboxylation - the formation of 3-phosphoglycerate;

- reduction - the formation of triose phosphates (glyceraldehyde 3-phosphate);

- regeneration - the re-formation of ribulose bisphosphate to accept more CO_2 molecules;

- product synthesis - the fate of triose phosphates.

Figure 2.8 describes all the reactions of carboxylation and reduction but it very much abbreviates the reactions of the regeneration phase. The reactions of the regenerative phase are sometimes likened to the reductive pentose phosphate pathway. The reactions of this pathway also feature in the catabolism of glucose where they function in the opposite direction; the oxidative direction. Do not forget, however, that the reactions of the reductive pathway are occurring in the chloroplast stroma, whereas those of the oxidative pathway take place in the cytosol. Therefore, they are independent of each other.

SAQ 2.5

We noted earlier that non-cyclic photophosphorylation produced NADPH and ATP in approximately equal quantities but we have just constructed an overall dark reaction which shows that 1.5 ATP are required per NADPH. Suggest a way in which the plant solves this problem.

2.5.2 Starch is synthesised in the chloroplast and sucrose is synthesised in the cytoplasm

Starch and sucrose are made from excess triose phosphate, and the pathway for their biosyntheses is shown in Figure 2.9. To follow this figure, begin with triose phosphate in the chloroplast stroma. In the chloroplast, follow its pathway to starch. Then beginning again with triose phosphate in the chloroplast stroma, follow its passage through the triose phosphate: Pi antiporter (transport system) into the cytosol and thence to sucrose.

triose phosphates

starch synthesis in stroma

The product of the Calvin cycle identified in Figure 2.8, glyceraldehyde 3-phosphate, is in equilibrium with dihydroxyacetone phosphate in a reaction catalysed by the ubiquitous enzyme triose phosphate isomerase. This is why these two metabolites are referred to as triose phosphates. In starch biosynthesis, two triose phosphate molecules combine to form the hexose, fructose 1,6-bisphosphate, which undergoes a series of changes leading to the formation of glucose 1-phosphate. This is converted into ADP-glucose which is the substrate for starch synthetase itself. This enzyme catalyses the transfer of a glucose molecule to a starch primer molecule which is extended by one glucose at a time.

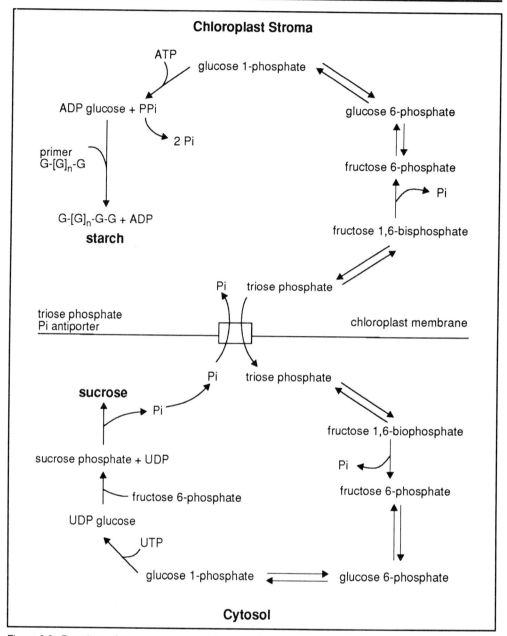

Figure 2.9 Reactions of starch and sucrose biosynthesis.

| **SAQ 2.6** | Can you think of anything we have forgotten to say about starch biosynthesis?

We remind you that starch consists of amylose (in which glucose is linked by α–1–4 links into linear chains) and amylopectin which is similar to amylose except it has a number of branch points in which glucose isomers are linked by α-1-6 linkage. |
| --- | --- |

In order to explain how amylopectin is made it has been postulated that two forms of starch synthetase are present, one forming the α 1-4 links and the other forming the α 1-6, which form the branch points. Recent evidence suggests that the 1-6 branch points might be formed by an enzyme which transfers the terminal glucose residue of the growing starch molecule to an internal glucose residue.

sucrose synthesis in cytosol

Sucrose is made in the cytosol from triose phosphate exported from the chloroplast. Reactions are carried out in the cytoplasm which lead to the formation of glucose 1-phosphate in just the same way as in starch biosynthesis. However, although the activities of the enzymes are the same, different isozymes are produced in the chloroplast and cytosol. After glucose 1-phosphate the pathways diverge; UDP glucose is formed by reaction with UTP and this reacts with a molecule of fructose 6-phosphate to form sucrose phosphate. A specific phosphatase cleaves the phosphate to produce sucrose.

The triose phosphate-Pi antiporter is a key element in the relations between the chloroplast and the cytosol. It comprises the astounding figure of 15% of the chloroplast envelope protein and functions to passively exchange Pi and triose phosphate. It can function only in a strict counter-exchange mode and this ensures that every triose phosphate molecule which leaves the chloroplast is counterbalanced by the entry of a molecule of Pi.

| SAQ 2.7 | Suggest two pieces of evidence which would support the idea that starch is generated in the chloroplast but not in the cytosol. |

The cell fractionation studies referred to in the response to SAQ 2.7 also show that sucrose phosphate synthetase and sucrose phosphate phosphatase are found only in the cytosol. Much is now known about the control of the processes of starch and sucrose biosynthesis but although most of it is beyond the scope of this chapter, we need to know a few key points because these will help us understand later chapters.

The key enzyme in starch biosynthesis is ADP glucose pyrophosphorylase which forms ADP-glucose and PPi (pyrophosphate). This is stimulated by 3-phosphoglycerate and inhibited by Pi. If sucrose biosynthesis is actively occurring, triose phosphate moves out into the cytosol thus reducing the stromal 3-phosphoglycerate concentration, and Pi moves in with the result that starch biosynthesis is reduced. Similarly, active starch biosynthesis reduces the export of triose phosphate and leads to an increased cytosolic Pi concentration. This indirectly leads to an inhibition of fructose 1,6-bisphosphate phosphatase which reduces sucrose biosynthesis. To some extent, therefore, starch and sucrose synthesis are competing processes. Furthermore, sucrose inhibits sucrose phosphate synthetase. Thus, a build up of sucrose for any reason causes feedback-inhibition of its synthesis and starch synthesis is increased.

| SAQ 2.8 | If leaves are sprayed with the mannose (a hexose), considerable quantities of mannose 6-phosphate are produced. However, starch production is increased and sucrose production decreased, when compared with control leaves. How can this result be explained? There is a hint in the response if you need it. |

2.6 Some plants produce a 4-carbon molecule as the first stable product of carbon reduction

We saw above that radiotracer experiments revealed that the 3-carbon compound 3-phosphoglycerate was the first stable compound formed in photosynthesis. Much of this early work was conducted using the green alga *Chlorella* and spinach. Following sound scientific principles, researchers extended the list of species tested and quickly

C3 and C4 plants

revealed that whereas many species of plant produce a 3-carbon compound, others, such as maize, produce a 4-carbon compound as the first stable compound. To distinguish easily between these two groups they are referred to as C3 or C4 plants.

C4 plants not only produce a 4-carbon compound but they also show a cyclical system of reactions in addition to the Calvin cycle. The detailed biochemistry of these reactions has now been elucidated and reveals that C4 plants can be divided into 3 sub-groups, which show slight metabolic variations from each other. The essence of these 3 schemes is the same, however, and is depicted in Figure 2.10. We have to discuss certain aspects of leaf structure before we explain this figure.

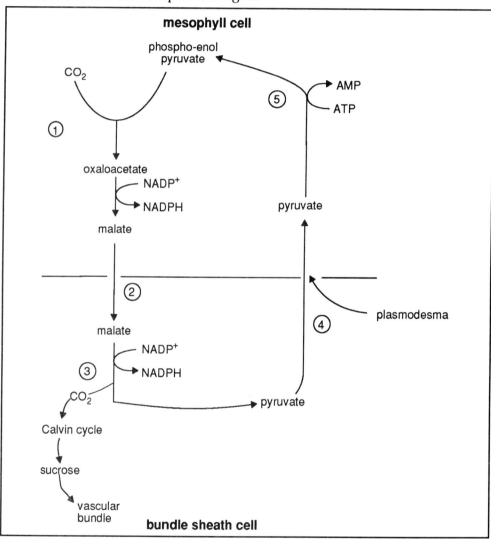

Figure 2.10 The fundamental C4 photosynthetic carbon assimilation cycle. The ringed numbers are referred to in SAQ 2.10. The enzymes of this cycle are described in the text.

The leaves of dicotyledonous C3 plants consist of palisade and spongy mesophyll cells, with vascular tissue and stomata arbitrarily distributed (Figure 2.11). The mesophyll cells contain chloroplasts and, therefore, are chlorenchyma. The chloroplasts are all of the same type.

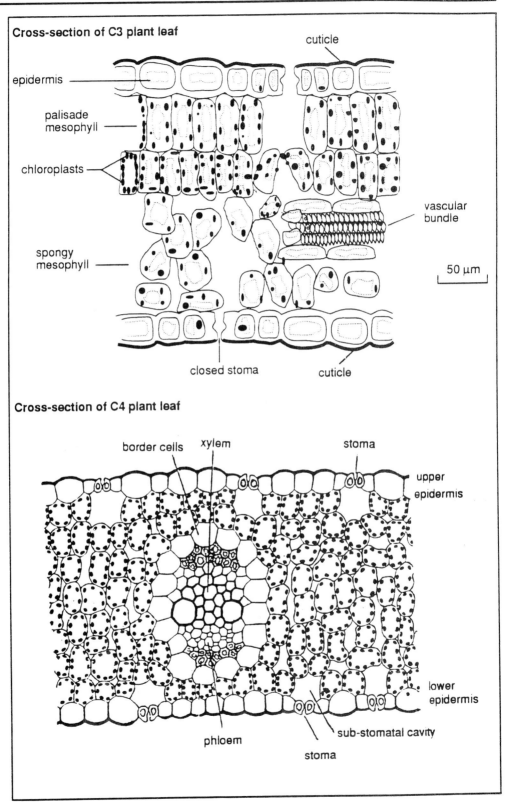

Fig 2.11 Stylised diagrams of cross-sections of typical C3 and C4 plant leaves.

This contrasts with the situation in C4 plants. Here the vascular tissue runs parallel to the long axis of the leaf and is surrounded by a sheath of cells, the bundle sheath. Two types of chloroplasts are evident; small chloroplasts containing grana in the mesophyll cells and very large ones lacking grana in the bundle sheath cells. Bundle sheath chloroplasts show low PSII activity and, therefore, release little O_2. Usually the stomata are arranged in lines parallel to the vascular tissue but are largely restricted to the spaces between the bundles; few stomata are found in the epidermis over the vascular bundles. Each stoma opens to a large air space, the substomatal cavity (SSC), but the mesophyll itself is tightly packed with little or no air spaces.

Many, but not all, monocotyledonous plants are C4 plants. C3 monocotyledonous plants have parallel veins but they do not have bundle sheaths and all their chloroplasts are of the same type.

We can now return to Figure 2.10. CO_2 diffuses into the leaf into the SSC and from there into the adjacent mesophyll cells. In the mesophyll cells, it reacts with phosphoenol pyruvate (PEP) in a reaction catalysed by PEP carboxylase, forming oxaloacetate which is then reduced to malate by the action of NADP-dependent malate dehydrogenase. Malate then diffuses through the symplasm into the bundle sheath cells, where NADP-dependent malic enzyme converts it into CO_2 and pyruvate, ie decarboxylation. The CO_2 is utilised in the Calvin cycle, the enzymes of which are found only in the bundle sheath chloroplasts, and the pyruvate then migrates back to the mesophyll. Here the enzyme pyruvate orthophosphate dikinase (POPDK) catalyses its conversion back into PEP. Thus, the function of this cycle is to transport CO_2 into the bundle sheath cells. You might like to write the names of these enzymes onto Figure 2.10.

SAQ 2.9

In the description of Figure 2.10, it was stated that malate diffuses into the bundle sheath cell and pyruvate diffuses back out. Bearing in mind that PEP carboxylase and POPDK are found only in mesophyll cells and malic enzyme only in the bundle sheath cells what is the driving force for these migrations?

SAQ 2.10

Figure 2.10 contains some numbers in circles, which refer to stages in the cycle. These stages are called transport, regeneration, decarboxylation, assimilation, but which number refers to which stage? One stage must be used twice to provide 5 answers.

It was mentioned earlier that C4 plants show 3 different schemes for the assimilation and metabolism of the primary C4 product. These schemes all contain the four stages referred to in SAQ 2.10, in the same sequence; they differ only in enzyme location and choice which need not concern us. Note that whereas NADPH is consumed in the assimilation stage, NADPH is reformed in the decarboxylation stage. Thus, there is no net consumption of NADPH. However, ATP is consumed in the regeneration phase and this constitutes a real cost to the plant. Moreover, 2 ATP equivalents are consumed because the regeneration reaction releases AMP. Thus, every molecule of CO_2 delivered to the bundle sheath cell costs 2ATP. Adding this to that used in the Calvin cycle means that 5ATP and 2NADPH are required per CO_2 fixed. In this sense C3 plants may appear more efficient than C4 plants but we will see below that we might wish to reconsider this statement.

2.7 C4 plants show higher rates of photosynthesis than C3 plants

Photosynthesis can be measured in a number of ways and popular methods involve the determination of the rate of CO_2 assimilation or O_2 release. Both of these methods give slightly inaccurate results because of the O_2 consumption and CO_2 release in respiration. However, the inaccuracy is indeed slight because under good light conditions photosynthesis is proceeding at 10-30 times the rate of respiration.

Table 2.1 shows that C4 plants typically carry out photosynthesis at twice the rate of C3 plants. How is this achieved? To answer this we must return to biochemical aspects of photosynthesis; specifically to the action of the first enzyme, in the Calvin cycle, ribulose 1,5-bisphosphate carboxylase. We will explain what the CO_2 compensation point means in Section 2.9.

	C4	C3
Rate of photosynthesis	$50\ \mu mol\ CO_2\ m^{-2}s^{-1}$	$25 \mu mol\ CO_2\ m^{-2}\ s^{-1}$
CO_2 compensation point	$5\mu l\ l^{-1}$	$50\mu l\ l^{-1}$

Table 2.1 Comparison of typical rates of photosynthesis and CO_2 compensation point in C4 and C3 plants at ambient CO_2 levels.

2.8 Ribulose bisphosphate carboxylase can also act as an oxygenase

The carboxylase which catalyses the reaction of ribulose bisphosphate with CO_2 can also react with oxygen:

$$ribulose\ bisphosphate\ +\ CO_2 \longrightarrow 2\ (3{-}phosphoglycerate)$$

$$ribulose\ bisphosphate\ +\ O_2 \longrightarrow 3{-}phosphoglycerate$$
$$+$$
$$2{-}phosphoglycolate$$

Rubisco The name of the enzyme is ribulose 1,5-bisphosphate carboxylase/oxygenase; commonly abbreviated to RuBisCO (or Rubisco).

Whereas the 3-phosphoglycerate is further transformed by the Calvin cycle enzymes this does not happen to the 2-phosphoglycolate. Instead, this compound undergoes a series of reactions (Figure 2.12) involving three organelles.

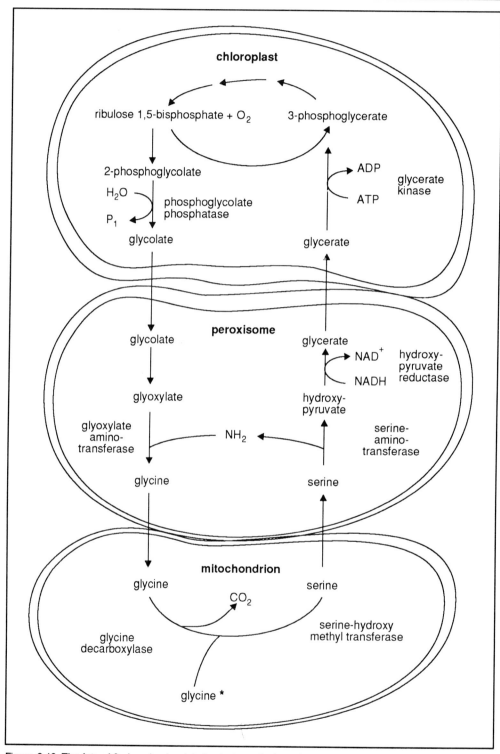

Figure 2.12 The fate of 2-phosphoglycolate in photorespiration, showing the involvement of three organelles (chloroplasts, peroxisomes and mitochondria). *This glycine is also derived from 2-phosphoglycolate produced by the oxygenase activity of Rubisco in the chloroplast.

There are several important things to note about Figure 2.12. It is, first and foremost, a scheme to scavenge some useful product from a larger amount of a virtually useless compound. To follow the figure begin with phosphoglycolate near the top and work your way round the cycle until you come to PGA (3-phosphoglycerate). Thus, two molecules of 2-phosphoglycolate are eventually converted into one molecule of 3-phosphoglycerate, which can enter the Calvin cycle. The cycle shown in Figure 2.12 involves the utilisation of O_2 and release of CO_2 and is, therefore, called the C2 photorespiratory carbon oxidation cycle (PCO), the overall phenomenon being known as photorespiration. The release of CO_2 constitutes a loss to the plant and reduces the net rate of photosynthesis.

PCO cycle photorespiration

∏ You might like to draw out the cycle shown in Figure 2.12 for yourself including the structures of the compounds involved. You should find any that you are not familiar with in standard biochemical texts.

peroxisome

The cycle, which recovers three quarters of the carbons in 2-phosphoglycolate, involves three organelles, as shown in Figure 2.12. (Note the cycle involves 2 glycine molecules derived from 2 phosphoglycolates). You should be familiar with chloroplasts and the mitochondria but may not yet have met peroxisomes. These organelles carry out a number of important metabolic reactions but usually does not show any particular internal membrane structure. They are often found in close proximity to chloroplasts and mitochondria. They are surrounded by a membrane and contain high levels of particular enzymes. (You should be able to list some from Figure 2.12).

The enzyme catalase, which degrades hydrogen peroxide, is found in such high concentration in peroxisomes that it becomes crystalline. This is very often used as a diagnostic tool to identify peroxisomes. The completion of the C2 PCO cycle, therefore, requires the cooperation of these three organelles.

SAQ 2.11

It would not be unreasonable to assume that a molecule of 3-phosphoglycerate could generate 18 ATP if it was fully oxidised through the glycolytic and tricarboxylic acid pathways. What energetic considerations should be borne in mind when proposing schemes to explain the transport of the various components between the three organelles which function in the C2 PCO cycle?

Experiments with the C3 plant sunflower show that at ambient partial pressures of CO_2 and 25° C the release of CO_2 from photorespiration is 20-25% of the rate of net photosynthesis. Figure 2.12 shows that two O_2 are utilised per CO_2 released in the PCO cycle so oxygenation must proceed in air at 25°C at 40-50% the rate of net photosynthesis. Thus, the ratio of carboxylation to oxygenation is calculated as about 2.5. The balance between the two systems is governed by the kinetic properties of RuBisCO and the concentration of CO_2 and O_2. Temperature has a double role. As the temperature increases oxygenation is enhanced relative to carboxylation. Further, as temperature increases the solubility of CO_2 falls more than that of O_2 so increased temperature reduces the CO_2:O_2 ratio in the cytosol. You may be asking what has all this to do with the higher rate of photosynthesis shown by C4 plants? We will now examine this.

2.9 C4 plants have a CO$_2$ concentrating mechanism

Figure 2.10 shows what we described as the CO$_2$ delivery system of C4 plants. This system also concentrates CO$_2$ in the bundle sheath cell thereby increasing the CO$_2$:O$_2$ ratio and reducing the oxygenase function of Rubisco found in these cells to relatively low activity. This is so efficient that C4 plants do not show photorespiratory loss of CO$_2$ and, therefore, typically show higher rates of photosynthesis than C3 plants. This lack of photorespiration also explains the differences in CO$_2$ compensation point, shown in Table 2.1. The CO$_2$ compensation point refers to the level of CO$_2$ at which the rates of photosynthesis and respiration are the same. It is measured by sealing a plant into a closed perspex chamber and allowing the CO$_2$ concentration to equilibrate. C3 plants typically maintain a value of approximately 50 µl l^{-1} whereas the value with C4 plants is approximately one tenth of this.

| SAQ 2.12 | What would happen to a C3 plant if it was incubated with a C4 plant in a sealed perspex container? |

We have previously noted that C3 plants could be considered to be more efficient than C4 plants because they use less ATP for every CO$_2$ fixed. However, Table 2.1 shows that C4 plants may carry out photosynthesis at double the rate of C3 plants.

How can this situation be explained? An obvious consideration is that since we have not defined efficiency we cannot properly compare the two strategies of photosynthesis.

C4 plants can photosynthesise more rapidly than C3 plants because they have evolved a CO$_2$ concentrating mechanism which enables them to avoid photorespiration. They avoid photorespiration by 'spending' more ATP, and thus could be said to be less energy efficient. But does this matter? Consider where the ATP comes from. It comes from the light reactions; ie. light is the driving force for making ATP and since light is free the extra ATP used by C4 plants does not constitute a loss.

This situation can be restated in the following way. C4 plants photosynthesise more rapidly than C3 plants because they have a CO$_2$ concentrating mechanism. This involves an apparently less efficient system of CO$_2$ fixation than in C3 plants because more ATP is used per CO$_2$ fixed. However, because ATP is obtained from light which is freely available and abundant, the lower efficiency of the partial reaction, ie. CO$_2$ fixation, is over-ridden and the completed reaction ie. photosynthesis itself, becomes more efficient.

2.10 CAM plants behave similarly to C4 plants

CAM plants

CAM is an acronym for Crassulacean Acid Metabolism. For many years this was thought to be a unique series of reactions shown only by members of the family Crassulaceae. We know now that the reactions are shown by many other Angiosperm families and, with the discovery of the CO$_2$ transport schemes of C4 plants, that the reactions themselves are not unique. The reactions referred to are shown in Figure 2.13.

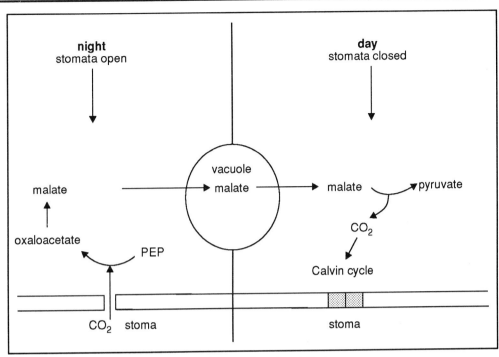

Figure 2.13 Crassulacean acid metabolism (see text for a description).

dark fixation of
CO₂

Members of the family Crassulaceae are unusual in that they close their stomata in the light and open them in the dark, which is the opposite to that shown by the majority of other plants. During the night, CO_2 is assimilated by the carboxylation of PEP to form oxaloacetate. This is reduced to malate which is stored in the vacuole. During the daytime malate returns to the cytosol where it is decarboxylated to produce pyruvate and CO_2, the latter being used in the Calvin cycle. Note that the stomata are closed during the day, thus significantly reducing any leakage of CO_2 or water vapour. The nocturnal acidification and daily de-acidification of these plants was clearly established in the 19th century but for many years was considered to be a biochemical oddity. The photosynthetic implications were not appreciated until the 1950s and the elucidation of the full details was stimulated by analogy with the C4 pathway which was discovered in the mid 1960s. We can describe the C4 pathway as one in which primary and secondary carboxylations occur, ie in the formation of the C4 acid and in the utilisation of the CO_2 released from it. C4 plants show a spatial separation of these two events and they possess distinct anatomical specialisations which facilitate this. CAM plants also show these two carboxylations but here they are separated in time. They do not show the anatomical specialisations of C4 plants but clearly possess biochemical specialisation in which, for example, the activity of key enzymes varies diurnally.

2.11 Stomata are the gateways for CO2 entry

Stomatal
control of
photosynthesis

The CO_2 which plants need for photosynthesis diffuses into the leaf mainly through stomata and as long as stomata are open plants are able to obtain CO_2. If the stomata are closed, CO_2 absorption is reduced by approximately 95%. Observations of stomatal behaviour of typical C3 and C4 plants reveal the results shown in Table 2.2.

Condition	Stomatal aperture
light	increases - stomata open
dark	decreases - stomata close
high CO_2 concentration	decreases - stomata close
low CO_2 concentration	increases - stomata open

Table 2.2 Effect of light and CO_2 concentration on stomatal behaviour.

In Table 2.2 high and low CO_2 concentrations refer to levels above or below normal (0.035%) in the intercellular air spaces near the stomata. Further to the results shown, the effect of dark over-rides that of low CO_2 and the effect of high CO_2 over-rides that of light. We saw in Chapter 1 that guard cells are specialised in having walls of varying thickness. This results in them responding to changes in cell turgor such that the stomatal aperture increases when guard cells become more turgid and decreases when the guard cells wilt. The shape of the guard cells is such that the stomata are probably never completely closed, there always being an open narrow channel, however contorted, running between them. However, it is convenient and common usage to refer to stomata opening or closing and that will be followed here.

We have emphasised that CO_2 diffuses into leaves via stomata but we must realise that water which evaporates from cells into the air spaces can also diffuse out. As long as water absorption by the plant roots matches that lost by evaporation from leaves, the stomata will remain open and photosynthesis will proceed. If water absorption is less than evaporation the leaves wilt and photosynthesis will be severely curtailed. For this reason, water availability controls leaf photosynthesis and is an extremely important factor affecting plant distribution.

Clearly, stomatal function depends indirectly on water absorption and transport and, therefore, so too does photosynthesis. In the next chapter we will examine cell and plant water relations and then return to discuss the mechanism of stomatal opening and closing in more detail.

SAQ 2.13

Bearing in mind what we have just said about the effects of water availability on photosynthesis, see if you can explain the effects of light and CO_2 on stomata.

Summary and objectives

This chapter began by stressing the importance of photosynthesis and described early work which led to a statement of the overall reaction and then of the light and dark reactions. The light reactions were described so as to explain the formation of ATP and NADP, by describing the operation of PSI and PSII and the links between them, and the function and structure of the light harvesting, antenna pigment systems. The description of the dark reactions began with the Calvin cycle in C3 plants and was followed by a description of the biosynthesis of sucrose and starch. The biochemistry and anatomy of C4 plants was described and an account given of photorespiration and its dependence on the carboxylase and oxygenase activities of RuBisCO. The role of the CO_2 concentrating mechanism of C4 plants in preventing photorespiration was stressed and this was linked to a discussion of CAM plants. Finally, the relationship between photosynthesis and stomatal operation was introduced and the dependence of both on water availability emphasised.

Now that you have completed this chapter you should now be able to:

- explain the nature of the interdependence of the light and dark reactions;

- demonstrate an understanding of the operation of electron flow and how the arrangement of components in the thylakoid membrane facilitates ATP production;

- calculate the ATP and NADPH requirements of the dark reaction and account for the number of quanta needed;

- describe some of the elements which control the production of sucrose and starch;

- describe an example of the function of the symplasm in intercellular transport;

- describe the biochemical and anatomical specialisations of C4 and CAM plants;

- show an understanding of the significance of the effects of light and CO_2 on stomatal behaviour.

Plants and water

Plants and water

3.1 Introduction

Terrestrial plants are considered to have evolved from aquatic plants and, therefore, inhabit an environment which is potentially hostile, in that water may not be freely or continuously available. Land animals, of course, are in a similar situation but animals can react to their environment by behaviourial traits which involve, for example, living near to or seeking out water. Plants have no such powers of locomotion and have evolved different strategies for survival. We saw in the previous chapter that plants are essentially unique in being able to carry out photosynthesis, but even this essential process takes second place to water conservation. Water is extremely important to plants. Plant vacuoles occupy between 80 and 90% of the plant cell volume and the content of these vacuoles is mainly water. Thus, it is not surprising to find that 90-95% of the mass of a plant is water. This water is continually on the move within the plant and we will see in later chapters that it has very important roles in transport. We have seen that it is used directly as a reactant in photosynthesis but also has indirect effects on this process through its effect on the behaviour of stomata. In this chapter we will study the basic elements of cellular water relations. We will apply these first to the mechanism of stomatal function and then extend the cellular aspects of water relations to the whole plant. Finally, we will examine a number of strategies employed by plants to aid water conservation.

3.2 Water distribution in plants involves water potential

water potential
ψ

A major factor governing the distribution of water in any system is the water potential of various parts of that system. Water potential is a measure of the ability of water to do work and is represented by the Greek letter psi (ψ). Thus, water will tend to diffuse from a region of high water potential to one of low potential, i.e down a water potential gradient. This movement will continue until the water potential is equal in all parts of the system. The diffusion of water under the influence of water potential gradients applies to the movement of water between plant cells and also between the plant and its environment. An important point to remember is that chemical potential and, therefore, water potential, is a relative term; it is always expressed as the difference between the potential of the substance in a given state and that in a standard state. The reference state of water is taken as that of pure water at ambient pressure and at the same temperature as the system under investigation. The water potential of this standard is arbitrarily assigned a value of 0 MPa, which, to some extent, can be considered to be unfortunate. As we will see, actual water potentials in plants and their environment are lower than this standard value and therefore, are negative. This does sometimes create potentially confusing situations, which would have been avoided if a value of, say, 500 MPa, had been chosen for the standard.

Several factors affect water potential, including solute concentration, pressure and matric potential, which will be described in more detail below. Before proceeding to describe these, however, we need to have a word about units. Concentration is normally measured in mass per unit volume, whereas pressure is obviously measured in pressure units. Clearly we have to convert all our units to a single unit so that comparisons can be made. By convention the common unit is the pressure unit and the SI unit for pressure is the Pascal (Pa). During the history of plant physiology several different pressure units have been used as shown in Table 3.1.

1 atmosphere	=	14.7 pounds per square inch
	or	760 mm Hg (at sea level)
	or	1.013 bars
	or	101300 Pa

Table 3.1   Comparison of units for measuring pressure.

Because of the magnitude of the number of Pa normally encountered, it is more usual to use a unit referring to a million (mega) Pa, ie MPa.

Now we are ready to examine the factors which affect water potential.

3.3 Solutes reduce water potential

The addition of solutes to water reduces the ability of water to do work and the effect is proportional to the total concentration of solute particles. Thus, the effect is additive. The unit of concentration can be converted to an approximate pressure unit as follows:

$$\text{concentration C (mol l}^{-1}) = - \text{RTC (MPa)} \qquad \text{(E-3.1)}$$

where R is the gas constant, (0.0083143 litre MPa $mol^{-1}K^{-1}$) and T is the absolute temperature. Note if we were to strictly apply SI units concentration would be given in mol m^{-3} and R in m^3 MPa mol^{-1} K^{-1}. However, litres are a more workable unit of volume than m^3 for most systems and are most commonly applied in biological studies, so we will stick with this convention. When discussing water relations, the solute concentration is termed the osmotic potential (π).

osmotic
potential π

The above conversion is not entirely accurate for two reasons. First, the concentration which will give the most accurate conversion is mol l^{-1} solvent not mol l^{-1} solution. Note that the mol l^{-1} solvent is called the osmolality of the solution and, as solutions become more and more dilute, it becomes closer and closer to the value mol l^{-1} solution. Let us do a couple of calculations before describing the second reason.

∏   Calculate the osmotic potential of the following at 20°C:

1) 0.01 mol glucose l^{-1} (relative molecular mass 180);

2) 0.01 mol sucrose l^{-1} (relative molecular mass 342);

3) a solution containing 0.01 mol glucose l^{-1} and 0.01 mol l^{-1} sucrose.

At 20°C, T = 293°K, thus:

RT = 0.0083143 x 293 = 2.436 (1 MPa mol^{-1}).

The osmotic potentials are:

1)   -0.01 x 2.436 = 0.0244 MPa;

2)   -0.01 x 2.436 = - 0.0244 MPa;

3)   - [(0.01 x 2.436) + (0.01 x 2.436)] = - 0.0488 MPa.

Thus, although the sucrose molecule is almost twice the size of the glucose molecule, each molecule contributes the same amount as a molecule of glucose to osmotic potential.

The second reason that the relationship is only approximate is because some substances dissociate in solution to produce extra solute particles, each of which contributes equally to the osmotic potential. Thus, 0.01 mol l^{-1} NaCl would contain 0.02 mol l^{-1} of solute particles, due to dissociation. Na_2HPO_4, if completely dissociated, would form four particles from each molecule. In these cases the effects of dissociation can be accounted for, but what about a substance whose dissociation is dependent on pH, for example, such as an amino acid? Here it would be much more difficult to allow for dissociation. This is also true when mixtures of salts are used, with variable numbers of particles. Thus, it is not possible to be precise about osmotic potential values of complex mixtures, especially at high concentrations.

The Greek symbol pi (π) is used for osmotic potential and, because solutes reduce water potential, π is, by definition, negative.

3.4 Pressure affects water potential according to its sign

turgor

tension

pressure
potential P

If pressure is applied to a solution, its ability to do work increases, ie its water potential is raised. Negative pressure will do the opposite. If a plant cell absorbs water its volume increases and it exerts a force on its cell wall, thus developing a positive hydrostatic pressure, called turgor pressure. Negative hydrostatic pressure can also develop, predominantly in the xylem and this is referred to as tension. Pressure is symbolised by the letter P and it is measured in pressure units. Remember that the reference potential of pure water was measured at ambient pressure but we are referring here to **differences** from ambient. Thus, the P value for a beaker of pure water would be 0 MPa, although its absolute pressure would be 0.1013 MPa at sea level.

| SAQ 3.1 | What is the absolute pressure value and the P value of water exposed to a perfect vacuum? Is it possible to have a positive water potential in a system? |

3.5 Colloids also affect water potential

matric potential

Many particulate materials absorb water and swell as a result. Such particles are termed colloids. Water is attracted to the surface of colloidal particles and because this reduces the ability of water to do work it reduces ψ. Such particles occur frequently in soils and occasionally in plants, such as when proteins and starch are stored. Plant cell walls also adsorb water molecules and can be said to have colloidal properties. These types of effect are grouped together in what is called the matric potential and they all reduce the value of ψ. The Greek symbol tau (τ) is used to denote the matric potential. Effects of τ are very important in soils but they are less important in plants because they tend to be fairly constant and are not used by plants to cause changes in ψ. Thus we will say no more about τ at this stage .

3.6 Total water potential

It follows from the above that we can write an equation for water potential:

$$\psi = \pi + P$$

where π is the osmotic potential and P the pressure potential. In these terms π is negative by definition and P is entered according to its sign. It is important to point out the variability which exists among texts in the way this equation is written. It is sometimes written:

$$\psi t = \psi s + \psi P$$

referring to total, solute and pressure potentials respectively. In some texts, the term osmotic pressure is used instead of osmotic potential. This is a positive term simply referring to the sum total of solutes and is equal to - osmotic potential. It is slightly misleading because of the word pressure but does have value in certain situations.

ψ gradients

Since π and P are in MPa units, so too is ψ and to restate the principle; water tends to diffuse from a region of high ψ to a region of low ψ .

3.7 Osmosis is a special type of diffusion

osmosis

Whereas water in the soil may diffuse freely down water potential gradients we are concerned mainly with movement within the plant, which usually involves movement between cells. This involves passage through the plasma membranes in the process of osmosis. Osmosis is the diffusion of a solvent through a semi-permeable membrane, down a chemical potential gradient of the solvent. In plants, water is the solvent and the plasmalemma is the semi-permeable membrane. Thus, in plants, water will move between cells by osmosis if there are ψ differences between them. In the absence of any limiting factor the process will continue until there is no gradient of ψ (ie until the cells have a uniform ψ). As we have mentioned, plant cells can develop a pressure against their walls, termed turgor pressure, which will counteract the effects of solutes, and thus lead to the removal of the gradient of ψ. Let us examine some examples of this.

Consider an open beaker containing water with no solutes. P is zero and π is zero, therefore ψ is zero. We now add sucrose to the beaker to create a solution of 0.1 mol l^{-1} sucrose.

$\prod$ Check back to our earlier in-text activity and decide what the ψ value of the water is now.

The answer is - 0.244 MPa (ie 10 x the value of our earlier in-text activity).

Consider also a flaccid, wilted cell containing solutes at 0.3 mol l^{-1}.

$\prod$ What is the ψ value of the cell?

The answer is - 0.732 MPa because P=0 and π is 3 x that of the sucrose solution in the example above.

$\prod$ What happens when we put the cell into the beaker of water? First of all decide if there will be any movement of water between the solution and the cell.

The answer here is yes, because the ψ of the cell at (- 0.732 MPa), is lower than that of the solution (- 0.244 MPa), water will move into the cell by osmosis, increase the cell protoplast volume and cause pressure to be exerted onto the cell wall. The cell wall, of course, responds with an equal and opposite pressure (ie turgor pressure = wall pressure).

$\prod$ Will the pressure continue to rise for ever or will it stop at some point? If it stops, explain why?

The pressure developing inside the cell will cause an increase in the cell's ψ until it equals that of the solution. At this point of equilibrium no further water entry occurs.

$\prod$ What will be the P value of the cell be at this point, assuming no change in the π value of the cell?

At equilibrium:

$$\psi_{cell} = - 0.244 \text{ MPa}$$

$$\pi_{cell} = - 0.732 \text{ MPa}$$

Since $\psi = \pi + P$:

$$P = \psi - \pi = - 0.244 - (- 0.732)$$

$$= - 0.244 + 0.732$$

$$= + 0.488 \text{ MPa}$$

It is important to realise that, at equilibrium, water is entering and leaving the cell at equal rates. Thus the water movement *per se* does not stop but net movement does. Further, we did the above calculation on the assumption that π cell was unchanged. This was not unreasonable because the entry of only very small amounts of water are needed before pressure begins to develop. This can be demonstrated by the calculation below.

cell wall rigidity   Wall rigidity is usually measured as a volumetric modulus of elasticity, that is the change in pressure (ΔP) divided by the relative change in volume ($\Delta V/V$). $\Delta V/V$ is the change in volume per unit volume.

$$\text{Elasticity} = \frac{\Delta P}{\Delta V/V}$$

Typical values for elasticity are 10 MPa, so if we suppose an increase in volume of 1%, (ie, $\Delta V/V = 1/100 = 0.01$) we can calculate that $\Delta P = 0.1$ MPa. A typical P value for a cell is 0.5 MPa so the 1% volume change, in causing a ΔP of 0.1 MPa, has increased P by 20%. A 1% volume change, would decrease, π by 1%. Thus, small increases in cell volume significantly alter P but not π.

Before proceeding further try the following SAQ.

SAQ 3.2
A flaccid cell with a π value of -0.2 MPa is placed in a large volume of solution with a ψ value of -0.12 MPa. Will there be any water movement? If so, in which direction will it move and what will be the P value at equilibrium?

SAQ 3.3
The cell used in SAQ 3.2 is now placed in a solution with a ψ value of -0.1 MPa. Will there be any movement of water and what will the equilibrium position be? Again, assume no change in π_{cell}.

Our cell, at the end of SAQs 3.2 and 3.3 is said to be turgid because it has a positive turgor pressure. If the cell is now placed in a solution with a ψ value of -0.26 MPa , water would diffuse out of the cell and the pressure would fall, eventually to zero. However, equilibrium would not have been achieved. You can see this from the following calculation:

$$\psi_{solution} = -0.26 \text{ MPa}$$

$$\pi_{cell} = -0.20 \text{ MPa}$$

If $P_{cell} = 0$:

$$\psi_{cell} = -0.20 - 0 = -0.20 \text{ MPa}$$

Thus, even when P is zero the water potential of the cell does not equal that of the solution.

plasmolysis   In our hypothetical example water would continue to move out reducing the protoplast volume so that it now no longer occupies all the space available within its cell walls. The plasmalemma shrinks away from the cell wall and the cell is said to show plasmolysis, (ie it is plasmolysed). Its appearance is shown in Figure 3.1.

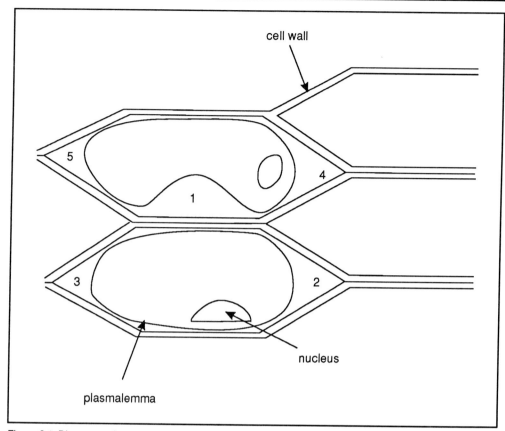

Figure 3.1 Diagrammatic representation of plasmolysed cells. Numbers are referred to in the text.

SAQ 3.4

This SAQ will test your understanding of points from Chapter 1 as well as from this chapter so far.

The cells shown in Figure 3.1 were produced by placing our hypothetical cells in a solution of $\psi = 0.26$ MPa. What is occupying the spaces indicated by numbers 1-5? Only one answer is expected here. The shape of the protoplast at 1 is quite different from elsewhere. Can you suggest an explanation for this type of plasmolysis?

<table>
<tr><td>SAQ 3.5</td><td>In an experiment, 25 chips were cut from a single large potato, each chip being 50 x 5 x 2 mm in dimension. The surface of each chip was carefully blotted with tissue paper and weighed in groups of five. These groups of five chips were incubated for 1 hour in sucrose solutions of various concentrations, blotted carefully and reweighed. The results of this experiment are shown below in Table 3.2.

What conclusions can be drawn from this experiment, within the context of water relations of the potato?</td></tr>
</table>

Chip group	Sucrose concentration (mol l^{-1})	Initial mass (g)	Final mass (g)
1	0.10	2.61	2.88
2	0.15	2.58	2.78
3	0.20	2.52	2.53
4	0.25	2.59	2.53
5	0.30	2.57	2.39

Table 3.2 Effect of sucrose concentration on change in mass of potato chips.

3.8 How are ψ, π and P measured?

It was pointed out above that approximate values of π can be calculated if the concentration of solutes is known, but that dissociation of salts and acids made it difficult to obtain precise values. This is one reason why a method of actually measuring π would be useful. Further to this, plant physiologists interested in water relations need to be able to determine all the parameters of the water potential equation and this can not be done by calculation. A number of methods are available to obtain the required data but we will examine only one, thermocouple psychometry. Psychometry estimates ψ by measuring the change in temperature during evaporation of water. It is based on the principle that evaporation of water from a surface cools that surface. The technique involves sealing a piece of tissue inside a chamber containing a temperature sensor, usually a thermocouple onto which has been placed a small drop of a solution or water. Evaporation occurs from both the drop of liquid and the tissue and the humidity of the air increases.

psychometry

If ψ_{tissue} is lower than ψ_{drop}, water will be absorbed by the tissue and continue to evaporate from the drop. Thus, its temperature will fall below ambient, in proportion to the difference in ψ between tissue and drop. If the solution has a lower ψ than the tissue, water will condense onto the sensor thus warming it above ambient. If the ψ of the tissue and drop are identical, a point is reached at which no further net evaporation occurs and the temperature of the drop is the same as the ambient temperature. The procedure is carried out several times coating the temperature sensor with solutions of known ψ and measuring the temperature after equilibrium. Figure 3.2 shows the type of result obtained.

By drawing the line of best fit through the data a point can be obtained at which there is no difference between sensor and ambient temperature. The solution causing this has the same ψ as the tissue and this can be read off the graph.

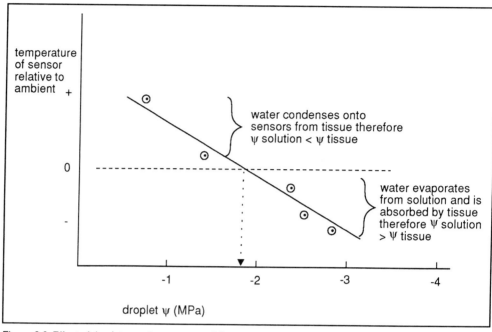

Figure 3.2 Effect of droplet ψ on temperature of the sensor in a thermocouple psychrometer.

The technique can also be used with a sample of cell sap instead of a piece of plant tissue. Thus, tissue ψ can be determined by using the tissue itself or by using samples of cell sap. The tissue P value can then be calculated. To this extent, the technique is very useful but it does depend upon very accurate temperature control of the bath in which the chamber is maintained. A temperature fluctuation of $0.01^{\circ}C$ can correspond to a ψ of 0.1 MPa so a very large water bath, usually around $1m^3$, is required to provide the necessary stability.

SAQ 3.6

See if you can identify another disadvantage to this technique, especially related to the calculation of P.

The problem raised in SAQ 3.6 is being approached by the development of direct pressure probe methods but at present these are limited to work with large cells due to the size of the probe itself. Techniques are also being developed which can be used in the field but these are beyond the scope of this text.

3.9 Stomatal mechanisms revisited

Now that we have examined water relations at the cellular level we can return to our discussion of the mechanism of opening and closing of stomata.

We noted in Chapter 2 that stomata open when guard cell turgor increases and close when they wilt. The material discussed above now allows us to explain this in terms of ψ. Let us begin with the guard cells wilted (P=0) and the stomata closed. If guard cell ψ now becomes reduced, water will move by osmosis from the cell walls surrounding the guard cells into the guard cell protoplasts. The protoplast volume will increase and a positive hydrostatic pressure will develop. Because of the shape of the guard cells and the manner of their wall thickening the two guard cells pull away from each other and the stomatal aperture opens. If guard cell ψ increases the opposite occurs and stomata close. As we saw above, only small volumes of water need to be moved to cause significant changes in water potential. The question now arises, what causes the changes in guard cell ψ?

Inherent to this answer are two further aspects of guard cell structure. First, guard cells are the only cells in the epidermis which may contain chloroplasts. Second, and uniquely, there are no plasmodesmatal links between guard cells and the cells which surround them, the subsidiary cells. Thus the guard cells are isolated from the symplasm of the epidermis. This is, perhaps, not surprising. The opening and closing of stomata depend upon very localised changes in turgor. If guard cells were part of the leaf symplasm any change in guard cell ψ would be dissipated because of the diffusive contacts with the surrounding cells. This phenomenon of symplastic isolation is used in plants whenever it needs to generate differences in ψ and π between adjacent cells and we will encounter it again in Chapter 5.

We will now examine the mechanism of stomatal opening and closing. It should be said at the outset that the full details of guard cell functioning are not yet known and that there are a number of unanswered questions. Plants manipulate guard cell ψ mainly by altering the distribution of K^+ ions between the guard cells and the apoplastic solution around them. The K^+ concentration in guard cells is 400 - 800 mmol l^{-1} when stomata are open and 50-100 mmol l^{-1} when closed. These changes in K^+ cause changes in π cell which, in turn, affect ψ cell. Thus ψ cell is controlled indirectly by the distribution of K^+ ions. This control is mediated by components situated on the guard cell plasmalemma and its directional aspect is affected by light, by CO_2 concentration around the guard cells and by abscisic acid (ABA), a plant hormone.

We will examine the action of light first. You will remember in the last chapter that ATP is generated in chloroplasts by utilising a proton gradient and the CF_0/CF_1 components of the ATPase in the thylakoid membrane. An ATPase is present on the guard cell plasmalemma but oriented in such a way that hydrolysis of ATP in the cytoplasm releases protons to the apoplast. Mitochondria generate ATP and light also provides ATP for the guard cell ATPase through the process of photophosphorylation. The protons released into the apoplasm not only reduce the apoplastic pH but also change the electrochemical potential across the membrane. This is a measure of the difference in charge between the inside and the outside of the membrane. The release of protons makes the potential more negative inside relative to the outside. This change in potential activates a K^+ channel, a component in the membrane which is specific for K^+ ions, and K^+ rapidly diffuses into the cytosol down its concentration gradient. Some evidence also suggests the presence of a K^+: proton co-transporter on the guard cell plasmalemma and that this is activated by light. Thus the action of light causes a rapid increase in guard cell K^+ concentration, which leads to an increase in the stomatal aperture. In darkness, K^+ moves from the guard cell into the apoplast and the stoma closes. Darkness does not appear to activate any particular process. Rather, in the absence of light, K^+ diffuses slowly out into the apoplasm and the stomatal aperture slowly decreases.

We noted above that CO_2 concentration and ABA (abscisic acid) affect guard cell functioning. In below-ambient CO_2 concentrations (in the light) stomata open and when CO_2 concentration is high they close. CO_2 also acts through changing K^+ levels but the details of the mechanism are not understood at present. The situation with regard to ABA is interesting. Whereas light causes rapid stomatal opening and its removal results in gradual closure, ABA works in the opposite way. When stomata are open in the light, the application of ABA causes rapid closure by causing a rapid release of K^+ from the cytosol. The evidence suggests that this is brought about by opening a diffusion channel in the membrane. Whether or not this is the same component which functions when light causes opening is not known. On removal of ABA, stomata slowly reopen. Thus, the actions of light, CO_2 and ABA all point to the central position of K^+ ion control of stomatal movement but much is still to be learned about the specific components which bring about K^+ distribution.

ABA causes stomatal closure

The situation is made more complex by at least three other points. The uptake of large numbers of positively charged ions drastically changes the balance of charges inside guard cells. This is rectified partly by the absorption of Cl^- ions, but mainly by the synthesis of malate. Phosphoenol pyruvate, (PEP), a monocarboxylic acid, is carboxylated to form oxaloacetate, a dicarboxylic acid, by PEP carboxylase. Oxaloacetic acid is reduced to malic acid by malate dehydrogenase, which dissociates to malate and hydrogen ions. The hydrogen ions are available to be pumped out by the ATPase, the malate accumulates and acts as the major counterion for K^+. Thus malate and K^+ ions both contribute to changes in guard cells π and thus ψ.

malate as a $2H^+$ donor

The second additional point concerns the fact that the guard cells of some species lack chloroplasts. The Ladies Slipper orchid, *Paphiopedilum* is an example. In such species, mitochondria provide the ATP for the ATPase. Thirdly, many species of plant possess a blue light sensitive system which also causes stomatal opening but it acts independently of photosynthesis. You should now be convinced that the apparently simple opening and closing of stomata is actually quite complex. Figure 3.3 shows a summary of these events.

Having discussed the mechanism of stomatal operation let us now restate the situation under consideration. Plants must photosynthesise to survive and for this they need CO_2. Since CO_2 enters via stomata plants need these to be open in order to obtain CO_2 but in addition to letting CO_2 in, open stomata will also allow water out. This does not present a problem as long as the plant can make up its water loss and it does this by absorbing water from the soil and transporting it up to the shoot. We will now examine this process.

stomata control gaseous exchange

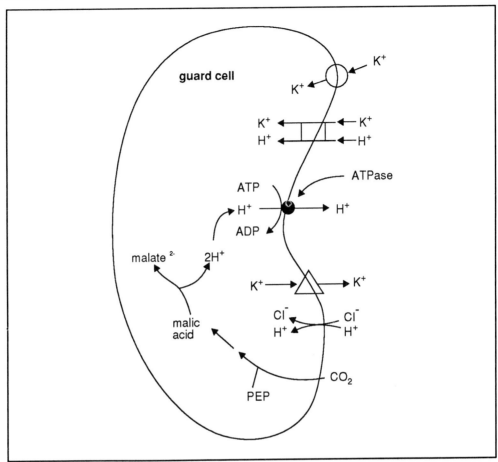

Figure 3.3 Summary of components proposed to be involved in stomatal opening and closure. O is the K^+ channel which allows K^+ to rapidly enter the guard cell when the light is on. The pumping out of protons raises the guard cell pH, which activates PEP carboxylase and the production of malate. Δ is the component which allows K^+ to rapidly move out of the guard cell under the influence of ABA. □ is a hypothetical K^+/H^+ cotransporter which is activated by light but inhibited by ABA. The components O and Δ may be a single identical component but at present this appears to be unlikely. Cl⁻ is considered to be absorbed by cotransporter with H^+ and to contribute in a small way to balancing the K^+ charge. It will do this, of course, only if the H^+, which is proposed to accompany its absorption, is removed by the ATPase.

3.10 Most water is absorbed by root hairs

The root has two functions, one is to anchor the plant so that it can hold its shoot upright in a position which allows light interception and the other is for water and mineral salt absorption. With regard to water absorption by roots there are two important considerations. Much of the water in soils is present as a coating of the soil particles. Soils, other than waterlogged soils, do not normally possess extensive tracts of free solution. Thus, effective water absorption by roots requires intimate contact between the surface of the root and the soil particles (Figure 3.4). Further, absorption will be in proportion to the root surface area making this contact. Plants solve this problem by the presence of their root hairs. Root hairs are fine, elongated extensions of epidermal cells, less than 0.5mm in diameter and approximately 5mm long. They are produced in the region of the root about 1 cm behind the growing point. They are continually being

function of root hairs

formed proximally and dying back distally to the tip and so maintain their position relative to the growing point.

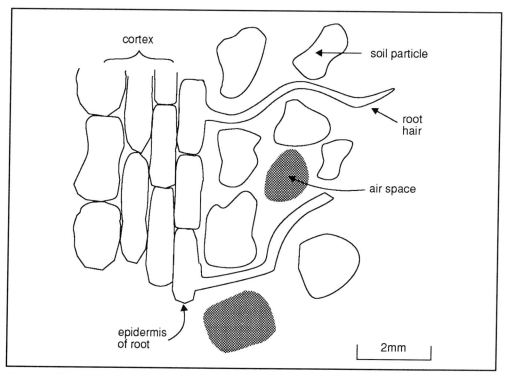

Figure 3.4 Diagram of root hairs extending between soil particles. Space in the soil not occupied by air spaces or soil particles is filled by the soil solution.

Root hairs have a high surface to volume ratio and can constitute up to 60% of the total surface area of a root. Under most conditions, the ψ of the soil solution is higher than that of the root hairs and cortical cells, so water will enter the root symplasm by osmosis. There tends to be a gradient of ψ across the root and so water will move radially inward until it reaches the xylem. The details of this process are described later in Section 3.15.

The water conducting xylem vessels and tracheids are dead, possess no plasmalemma and so do not exhibit osmosis. They form essentially long connected tubes, not totally open because of the persistent end walls, but partially open because of various wall perforations. Their structure suggests that the transport of water through them is by bulk flow, so let us now examine the forces needed to drive such movement.

3.11 What pressure gradient is required for bulk flow in the xylem?

The velocity of transport of water varies among species and depends not only on the composition of the xylem, ie whether vessels or tracheids are present, but also on the actual transpirational demand. Trees with numerous wide vessels typically show transport velocities between 4 and 14 mm s^{-1}, but this is reduced to between 0.3 and 1.7

mm s^{-1} in trees with narrower vessels. We will use a value of 5 mm s^{-1} and a vessel radius of 50 μm in our example.

The Poiseulle equation shows that:

$$\text{volume flow rate} = \frac{\pi\, r^4}{8\, \eta} \times \frac{\Delta P}{\Delta x} \qquad \text{(E-3.2)}$$

where r is the radius of the tube, η is viscosity and $\Delta P / \Delta x$ is the pressure gradient.

Dividing equation 3.2 by πr^2, the cross sectional area of the tube, we obtain:

$$\text{velocity of transport, } J_v = \frac{r^2}{8\, \eta} \times \frac{\Delta P}{\Delta x}$$

If we assume that the viscosity is similar to that of water, (10^{-3} Pa s^{-1}), a J_v of 5 mm s^{-1} through a vessel of 50 μm radius will give us:

$$5 \times 10^{-3}\, \text{m s}^{-1} = \frac{(50 \times 10^{-6})^2}{8 \times 10^{-3}\, \text{Pa s}^{-1}} \times \frac{\Delta P}{1\, \text{m}}$$

thus, $\Delta P = 0.016$ MPa m^{-1}.

How might such pressure gradients be generated in the plant? One of the early ideas to explain this was the generation of a positive force in the roots which pushed the fluid up. This was suggested because the roots do, indeed, generate a positive pressure. This can be demonstrated very easily by attaching a manometer to a stem cut just above soil level. Some plants, such as grapevines, show a considerable root pressure, often as high as 0.6 MPa.

∏ Let us calculate how high a column of water would be supported by such a force. The barometric height of mercury, the height supportable by a ΔP of 0.1 MPa (1 atmosphere), is 760mm. You probably knew this already. What you might not have known is the equivalent height for water. This is approximately 10m. You should now be able to answer the question.

The answer is 60m : a ΔP of 0.1 MPa supports 10m, therefore a ΔP of 0.6 MPa would support a water column of 60m.

Root pressure thus seems to be a quite reasonable candidate. However, for two reasons we no longer consider this to be the case.

root pressure   First, root pressures can only be demonstrated in well-watered plants. The moment there is any water deficit in the soil the plant loses its ability to generate root pressure. Second, it has not proven possible to demonstrate a root pressure in conifers. Thus, attractive as it is, this theory does not solve the problem.

Another idea proposed in the early days of plant physiology was that water was transported by capillarity, the upward movement of water in narrow tubes. The phenomenon was discovered using glass tubes but xylem vessels and tracheids also consist of narrow tubes. Calculation of the lifting force and the downward force and rearranging the equation shows that the relation between radius and height in capillarity is given by:

$$h = \frac{1.5 \times 10^{-6} \text{ m}}{r}$$

where r is the radius of the tube and h the height of the column, both in metres.

∏ Calculate h for the r values of 1, 10, 100, 1000, 75 and 40μm (without looking at the results below).

radius (μm)		height (m)
1		1.5000
10		0.1500
100		0.0150
1000		0.0015
typical vessel	(75)	0.0200
typical tracheid	(40)	0.0380

These calculations show that although very narrow tubes will support a column 1.5m high, typical xylem cells would manage only a fraction of that. On these grounds, capillarity alone cannot be important in moving water up trees.

transpiration pull

The current theory to account for water movement was proposed by Dixon and Joly, who suggested that water evaporating from the leaves (transpiration) was the driving force. This evaporation created a tension in the water columns in the xylem and the pressure gradient needed was achieved by a sufficiently negative internal pressure working against the external ambient pressure. Dixon and Joly noted that such a system would require very strong tubes, which could withstand the tension without collapsing, and the internal fluid would need a very high tensile strength, so that the columns of water would remain intact. Dixon and Joly suggested that the lignified xylem vessels and tracheids would have sufficient strength and tests showed that the tensile strength of water is very high (greater than -30 MPa).

cavitation

These measurements of water tensile strength reveal the importance of gas dissolved in the water. As the pressure is reduced, gases tend to come out of solution and since gases cannot withstand negative pressure, the formation of a gas bubble cause the collapse of a water column. This phenomenon is referred to as cavitation and we will see later that it does sometimes occur in nature.

Thus, the system seems to have the physical requirements needed for Dixon and Joly's theory but is there any evidence in favour of it? There are several pieces and we will examine some of them now.

3.12 Tension causes twigs and branches to shrink in diameter

If a healthy transpiring shoot is severed from its plant and the cut surface examined under a microscope it appears dry and dull. If the shoot is placed in a pressure chamber and the cut surface observed as pressure is applied, small droplets of water can be seen to emerge from the cut cells. This is a clear demonstration of the presence of tension within the plant. This was taken a stage further by MacDougall who predicted that if tensions did develop in transpiring plants this would, through the concerted action of all the cells in a cross section, reduce the diameter of twigs and small branches. He tested this by developing his dendrograph, an instrument for accurately measuring the circumference of branches. These tests showed that branch diameter did indeed, change; it was smallest when transpiration was occurring and largest when it was not. This is corroborative evidence for the occurrence of a tension in the xylem vessels and tracheids.

SAQ 3.7	Devise a method of directly measuring the pressure inside the xylem of a leaf twig.

3.13 Sap flow measurements corroborate the Dixon-Joly theory

Sap flow can be measured in plants by the use of heating coils and temperature sensors (Figure 3.5).

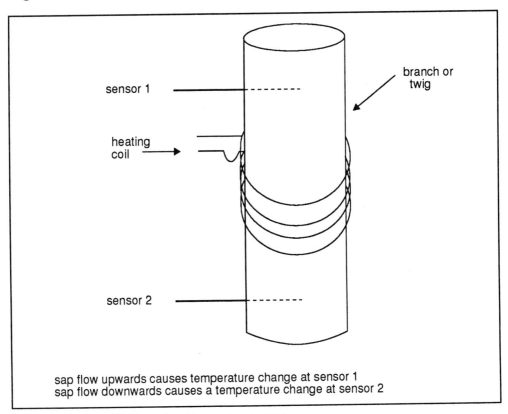

Figure 3.5 Diagram of heating coil and temperature sensors.

The heating coil is wound around the trunk, branch or twig and when switched on heats the sap inside the plant in the area surrounded by the coil. Small cores of tissue are removed above and below the coil and temperature sensors inserted. If sap is flowing, the direction of sap flow can be decided from which sensor detects a temperature change.

Let us now consider an experiment in which three of the heating coil and temperature sensor units of the type shown in Figure 3.5 were attached to a 5m tall tree. One was placed around the trunk, one around a branch and one around a thick twig (see A,B and C in Figure 3.6).

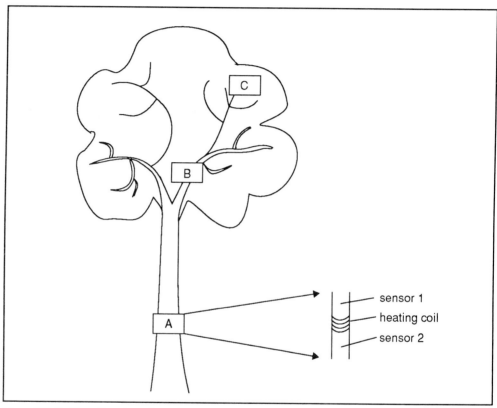

Figure 3.6 Position of heating coils and sensors in an experiment to test for sap flow in a tree. Sensors are placed on either side of the heating coils placed at A, B and C. Only those at A are depicted.

SAQ 3.8

The experiment shown in Figure 3.6 was set up on day one and the heating coils switched on at dawn on the next day. Now, make two predictions.

1)   Assume that root pressure is the driving force for water movement in the tree and that it begins to be generated as the sun comes up. Will sensor 1 or 2 at each unit signify sap movement and in what order will the units give positive results?

2)   Assume that evaporation of water from the leaves is the driving force for water movement and answer the same questions posed in 1).

When the experiment described in SAQ 3.8 was carried out, the order of detection of flow was C,B,A. This can be interpreted as showing that water evaporating from leaves created a tension which was then propagated down the tree until sap was flowing in all parts.

3.14 The problem of cavitation

cavitation

We have noted that considerable tension develops in the xylem and, because the water in the xylem contains dissolved gases, there is a tendency for gas bubbles to form. This is called cavitation and it breaks the continuity of the water column thus stopping water transport. This phenomenon does occur naturally and is accompanied by a sharp click which can be picked up by a sensitive acoustic receiver. Cavitation could theoretically be disastrous but plants survive it by several means. Once an air bubble is formed it tends to be limited to an individual vessel or tracheid element because surface tension prevents its spreading through the pits (see Chapter 1). The torus pits of conifers appear to function specifically to prevent the lateral spread of a bubble. Because the tracheids and vessels are connected laterally through pits and because the bubble tends to be localised, it is considered that water can flow around a cavitated element. Finally, at night transpiration slows or almost stops and the tension is dissipated. Thus the bubble may well go back into solution. This is even more likely to occur if the plant develops root pressure at this time.

3.15 What route does water take in the leaf?

The xylem in the stem delivers water to the leaf via the system of veins. These are usually between 3 and 10 cells away from the sub-stomatal air spaces which are linked to the outside world by the stomata. Water could follow two pathways between the vein and sub-stomatal space, an apoplastic and a symplastic pathway. We can use the same calculation here as in Section 3.11 above to calculate the pressure gradient needed for water to flow through the apoplast at 5 mm s^{-1}. If we make a very general assumption that the cross sectional area of the pathway is $2\mu m^2$, then the pressure gradient needed is 10 MPa m^{-1}. We can relate this to the pressure gradient needed for the symplastic route which involves passage through one membrane to enter the symplasm, passage through another to get out, and movement between the two. Passage though a membrane depends on L_p, the hydraulic conductivity, and the $\Delta\psi$ across the membrane:

$$J_v = L_p \times \Delta\psi$$

Using an L_p value of $4 \times 10^{-7} \text{ ms}^{-1} \text{ MPa}^{-1}$ and the same flow rate as before, (5mm s^{-1}) we have:

$$\Delta\psi = \frac{5 \times 10^{-3} \text{ ms}^{-1}}{4 \times 10^{-7} \text{ ms}^{-1}\text{MPa}^{-1}} = 1.25 \times 10^4 \text{ MPa}$$

This figure needs to be doubled because two membranes must be crossed, thus $\Delta\psi = 2.5 \times 10^4$ MPa. Let us assume that the symplastic distance is $100\mu m$ $(10^{-4}$ m), so the gradient becomes:

$$\frac{2.5 \ \times \ 10^4 \ \text{MPa}}{10^{-4} \ \text{m}} = 2.5 \ \times \ 10^8 \ \text{MPa m}^{-1}$$

The driving force needed to maintain a rate of water movement through cells is therefore, 2.5×10^8 MPa m^{-1}. This is considerably higher than that needed for movement through the apoplast (10 MPa m^{-1}). This would suggest that most of the water which passes from the vein to the outside would travel through the cell walls. This can be tested experimentally with a simple potometer. A potometer is a device which allows the rate of transpiration to be measured by observing the rate at which an air bubble moves along a capillary (Figure 3.7).

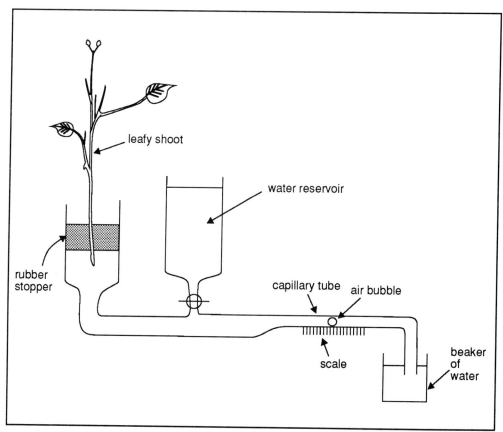

Figure 3.7 Diagram of a potometer.

If a leafy shoot is inserted into the instrument , water will evaporate from the leaf surfaces and create a tension in the system. The rate of transpiration is now measured by determining the rate at which the air bubble moves along the capillary tube. Transpiration is then stopped suddenly, by dipping the shoot in mineral oil, and the rate of movement of the bubble observed as the tension within the leaf is dissipated. Results of this type of experiment are shown in Figure 3.8.

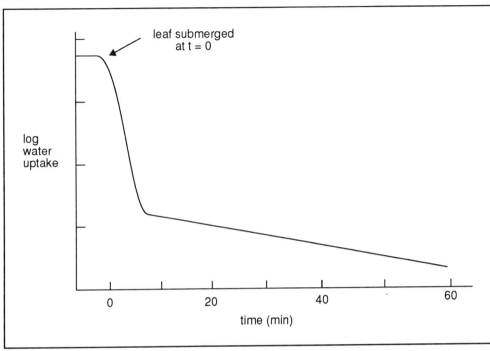

Figure 3.8 Decline of rate of water absorption during dissipation of a previously generated transpirational tension. Note the log scale of the rate of water uptake.

The results show that a large part of the tension is relieved very quickly but a second smaller portion is relieved only slowly. The fast reaction is interpreted as taking place in the apoplast and the slow reaction in the symplasm. Experiments of this type were first done by Weatherley in 1963 and they are interpreted as being consistent with our statement above, that the apoplast is the preferred pathway.

SAQ 3.9

What is the principal limitation of using a potometer to study plant: water relations?

3.16 The atmosphere has very low ψ values

To complete our account of the movement of water we must briefly consider the ψ of the atmosphere. We have described water evaporating from the leaf but this will only happen if there is a suitable gradient of ψ. As it happens, air has a very high capacity to hold water. This is usually referred to in terms of relatively humidity (RH), which represents the water content as a percentage of the maximum at a given temperature. These can be converted to ψ values (Table 3.3).

ψ and RH

Notice that water at RH 90% has a ψ of -32.8 MPa. When we realise that typical ψ values for leaves are in the order of - 0.8 MPa it becomes obvious that there is a considerable ψ gradient under all RH values other than those close to saturation. This being the case, how can we reconcile the fact that generally speaking transpiration is higher in moving air than in still air, even when the RH of the moving air is constant? This raises the question of resistance to water flow which we will now consider.

resistances to water flow

Relative humidity	Water potential (MPa)
99%	-3.12
90%	-32.80
50%	-215.50
20%	-500.00

Table 3.3 Relationship between relative humidity and water potential at 20°C.

3.17 A number of resistances to flow can be recognised

boundary layer

The situation referred to above, in which transpiration was higher in moving air, led to the recognition of what is called boundary layer resistance, caused by the presence of an unstirred layer of air next to the leaf surface. This layer acts as a buffer between the leaf and the bulk of the atmosphere and reduces water loss. Removal of the layer by blowing air over the leaves thus increases transpiration. This phenomenon allows the rationalisation of a number of plant strategies. Thus, the presence of hairs (trichomes) on leaves will increase the thickness of still air and increase the boundary layer resistance. Sunken stomata in which stomata are present at the base of small indentations or pits on the leaf serve the same purpose.

role of
trichomes and
sunken stomata

In the earlier part of this chapter we have accounted for transpiration in terms of $\Delta\psi$ but how does this explain why plants transpire less when the light is turned off? Of course the stomata will close in the dark, markedly reducing water movement through them and this effect is described as being due to stomatal resistance. When stomata are closed water loss occurs slowly through the cuticle, which is the site of cuticular resistance, a resistance which can be markedly increased by increasing the thickness of the cuticle (Figure 3.9).

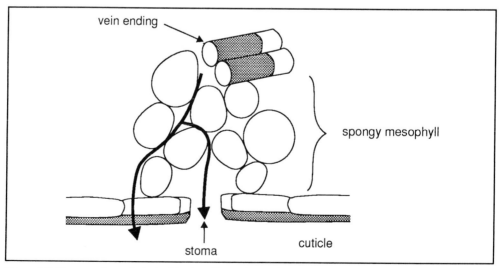

Figure 3.9 Diagram of a part of a leaf to show the pathway of movement of water from the vein ending to the exterior. The thickness of the mesophyll, the diameter of stomata and the thickness of the cuticle will all have an effect on the rate of water loss.

This concept of resistances to water movement is analogous to an electrical circuit, where the magnitude of the current is governed by the potential difference divided by the resistance. This analogy is useful because it allows us to explain many naturally-occurring water-flow phenomena. These resistances already mentioned plus several others are represented diagrammatically in Figure 3.10.

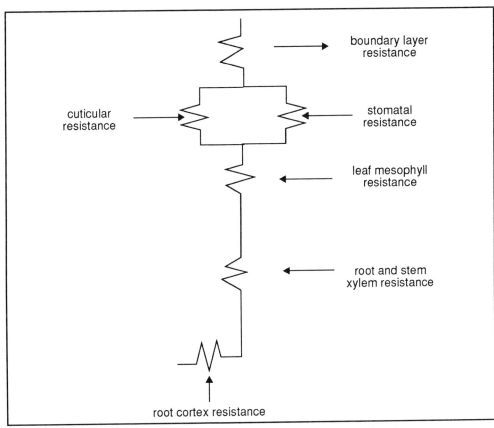

Figure 3.10 Diagrammatic representation of the resistances to water movement in a plant.

Let us test your understanding of these points with an SAQ.

SAQ 3.10

Explain the following:

1) The rate of water movement of 4 μl min^{-1} was recorded in a potometer containing a leafy seedling but this was reduced to 1.5 μl min^{-1} when the water in the potometer was replaced with 0.1 mol l^{-1} sucrose.

2) A rate of water movement of 3.5 μl min^{-1} was recorded with another seedling but when the roots were excised under water and the plant replaced in the potometer the rate observed was 5.0 μl min^{-1}.

3) Would you expect the same result in 2, if the plant roots were cut off in air rather than under water? Explain your answer.

4) Make a list of the morphological features you might find in a leaf which would indicate that the leaf was specialised to reduce water loss.

3.18 The role of abscisic acid in root to shoot water relations

Abscisic acid (ABA) is a plant hormone with a number of effects, an example of which is its ability to cause stomatal closure, as described in Section 3.9. Its structure is shown in Figure 3.11.

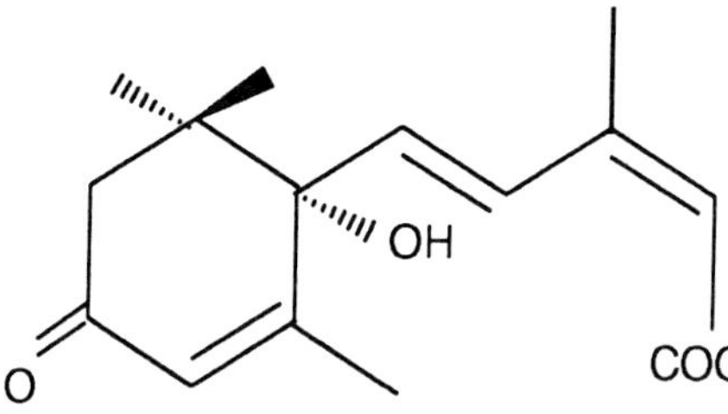

Figure 3.11 The structure of (+) -2 - cis abscisic acid.

When mesophyll cells wilt, ABA is synthesised and released into the apoplasm. It diffuses from there to the guard cells and brings about the effect we noted earlier. Thus, ABA can be looked upon as a signal that the leaves are short of water, the signal being transferred to the guard cells which react by causing stomatal closure. This reduces water loss whereupon the mesophyll cells regain turgor and switch off ABA synthesis. ABA already produced is gradually broken down, the guard cells slowly recover turgor and the stomata reopen. Remember, it is important for stomata to be open so that leaves can obtain CO_2 for photosynthesis.

root ABA controls leaf water loss

This effect of ABA has been known for many years but a very exciting extension of this story has recently been unravelled. When a soil dries out, water loss from the plant exceeds water absorption and the leaves wilt, produce ABA and close their stomata. However, a considerable water stress is required before leaves produce ABA and the roots have already experienced this stress before leaf ABA is produced and stomata closed. Recent experiments suggests that plant roots are able to signal to the shoot that they are experiencing water stress so that stomata are closed *before* the leaf experiences this stress. The signal is ABA, produced in the root cap as a result of water stress. This ABA travels in the transpiration stream to the leaves where it acts on the guard cells to cause closure. Many of these experiments involved the use of maize plants grown with their roots divided into two portions, each grown in a different pot; a split root design. It was found that if one pot was allowed to become dry the stomata on the leaves closed, even though the shoot obtained ample water from the other portion of root and showed no sign of wilting. Improvements in the assay procedures for ABA and in the micro-dissection of leaves was needed to show the presence of increased quantities of ABA in the apoplasm of the epidermis of such treated plants. These experiments suggest that the roots of maize produce ABA when water-stressed and use this to cause the shoot to close its stomata and thus reduce water loss.

| SAQ 3.11 | Which of the following would be necessary for the demonstration of the operation of the root to shoot signalling system just described? Explain you answer. |

1) Leaves showing wilting.

2) Leaves showing no wilting.

3) Enhanced ABA levels in leaf mesophyll.

4) No increase in ABA level in mesophyll.

5) Production of increased ABA levels in the root.

The experiments using split roots show that ABA synthesis indeed occurs in the dried roots but the leaves of the plant do not wilt and neither do the mesophyll cells produce ABA under these conditions. Scientists are presently testing a wide range of species to determine how widespread this phenomenon is.

3.19 Is transpiration a necessary evil?

evaporative coolings and mineral transport

Does transpiration have any function or is it nothing more than the price that plants pay for carrying out photosynthesis? In actual fact it has two clear functions. Evaporation of water cools the surface from which it occurs and this is very important in dissipating the plants' radiation load. Thus, as long as plants have access to sufficient soil water so that they can continue to transpire they have the ability to counteract the increase in temperature which light absorption causes. Because of this phenomenon, plant leaves are normally slightly below ambient temperature and are cool to touch. This is obviously not the case for plastic leaves and flowers, and surreptitiously checking if they are cool to touch can often permit the avoidance of embarrassment at smelling a bunch of artificial flowers! Thus, heat loss is a very important function of transpiration. A second function is mineral transport. Since plants obtain their minerals from the soil, they must be able to transport them around the plant. They do this by utilising the transpiration stream as a delivery system and that provides us with the link to the next chapter.

Summary and Objectives

In this chapter we have discussed the water loss problems which plants suffer as a result of retaining open stomata for photosynthesis and describes how plants overcome these problems. Water distribution was explained initially in terms of water potential gradients and the effect of solutes, pressure and matric potential were described. An outline of the process of osmosis, the generation of turgor pressure and the phenomenon of plasmolysis was given. This information was then applied to the operation of stomata and the role of potassium was explained. The involvement of osmosis in whole plant water relations was discussed and the possible contribution of capillarity and root pressure described. Long distance transport was explained by the Dixon-Joly theory and some evidence for it examined. The role of resistances to water movement was raised and possible resistances identified to explain the effects of thick cuticles, hairy leaves and succulent leaves. We concluded the chapter by discussing the role of ABA in root to shoot water relations and the possible functions of transpiration.

Now that you have completed this chapter, you should be able to:

- describe the components of the water potential equation and solve cellular water relations problems from given data;

- describe how ψ, π and P may be measured;

- describe the role of potassium ions in stomatal function and suggest mechanisms for the action of light, CO_2 and ABA;

- calculate pressure gradients needed to account for given data on the bulk flow through the xylem;

- show an understanding of the contribution of $\Delta \psi$ and resistance to flow in whole plant water relations by analysis of provided data;

- identify two functions of transpiration and describe a role for ABA in root to shoot water relations.

Mineral nutrition and transport

Mineral nutrition and transport

4.1 Introduction

We noted in earlier chapters that terrestrial plants face the problem of a somewhat hostile environment and must possess strategies which allow them to withstand this hostility. They also face a problem regarding mineral nutrition. In an aquatic environment, the whole plant is surrounded by water containing mineral salts so there is no need for a transport system. In a terrestrial plant, however, the roots may have direct access to minerals but the shoot does not. Thus, there is a need for a mineral transport system. We saw in the previous chapter that minerals are transported in the transpiration stream. We will now investigate how minerals get into the transpiration stream, where they are transported to and what their functions are when they get there.

4.2 Aristotle considered that plants eat soil

As with many early Greek philosophers, Aristotle had opinions about how plants grew and considered that they somehow took the soil into themselves and were actually made of soil. Anyone who has examined the root system of an extremely pot-bound plant to find such a large amount of root and small amount of soil might be tempted to agree with Aristotle. Indeed, this view prevailed until van Helmont carried out his celebrated experiment in the early 17th century. He planted a small willow tree weighing 2.2kg in 90kg of soil and did nothing other than water it with rain water over the next five years. At the end of this period the tree weighed 76kg but the soil weighed only 50g less. So van Helmont concluded that plants are made from water. The first clue to the fact that Aristotle and van Helmont were both partially correct was obtained by Woodward, towards the end of the 17th century. Woodward can be considered to be the father of hydroponics for he was the first to study the growth of plants with their roots in water. He used garden mint plants and showed that they could survive for some time in rain water but they were much healthier in water from the river Thames. However, they grew best of all in a watery extract of soil.

4.3 Hydroponics allows us to decide which minerals are essential

hydroponics   Hydroponics consists of growing plants with their roots immersed in a solution of mineral salts (Figure 4.1).

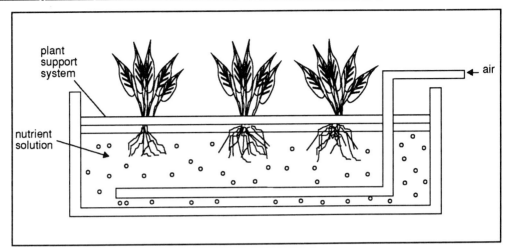

Figure 4.1 Hydroponic growth system.

The tank is partially sealed to prevent water loss and the plants must be supported because they lack the normal anchorage provided by the roots.

SAQ 4.1

Plants will grow much better in hydroponics if the solution is aerated. Explain why this is so.

essential elements

Since we are able to make up known mixtures of mineral salts we can use hydroponics to decide whether or not an element is essential. Three criteria are used to decide upon essentiality. First, the element must be required for normal growth of a plant and for completion of its life cycle; neither must occur in the absence of the element. Second, the requirement for the element must be specific and not replaceable by a different element. Third, the element must be acting directly in the plant. It should not act by aiding the absorption of another element nor by antagonising the toxic effects of other elements. As you can see, these criteria are quite strict and sometimes need to be modified.

Applying these rules to the process of hydroponics we are now in a position to be able to specify which elements are essential. This list is shown in Table 4.1.

You will realise that the large number of mineral elements are required by plants (we list sixteen of these in Table 4.1). You should try to remember these. Those that are required in larger amounts (the nine at the bottom of the table) are frequently referred to as macro-elements. These are the easier ones to remember since you can probably think of their roles in cells. The micro-elements (the seven at the top of the table) are more difficult. However, with a little practice, you should be able to recall them all.

Element	Chemical symbol	Form available to plants	Concentration in dry tissue (μmol g^{-1})	Relative No. of atoms compared to molybdenum
molybdenum	Mo	MoO_4^{2-}	0.001	1
copper	Cu	Cu^+, Cu^{2+}	0.01	100
zinc	Zn	Zn^{2+}	0.3	300
manganese	Mn	Mn^{2+}	1.0	1 000
iron	Fe	Fe^{3+}, Fe^{2+}	2.0	2 000
boron	B	$BO_3^{2-}, B_4O_7^{2-}$	2.0	2 000
chlorine	Cl	Cl^-	3.0	3 000
sulphur	S	SO_4^{2-}	30	30 000
phosphorus	P	HPO_4^{2-}	60	60 000
magnesium	Mg	Mg^{2+}	80	80 000
calcium	Ca	Ca^{2+}	125	125 000
potassium	K	K^+	250	250 000
nitrogen	N	NO_3^-, NH_4^+	1000	1 000 000
oxygen	O	O_2, H_2O	30 000	30 000 000
carbon	C	CO_2	40 000	35 000 000
hydrogen	H	H_2O	60 000	60 000 000

Table 4.1 Form of mineral absorbed and concentration considered adequate in tissue. (After Epstein, E Mineral Nutrition of Plants, Wiley, Chichester, 1972).

micro- and macro-elements

The table is arranged in ascending order of concentration considered adequate. This shows a very striking sequence. Note for example that a million times as much nitrogen is required compared to molybdenum. Table 4.1 shows that the elements fall into two natural groups. Molybdenum down to chlorine are referred to as micro-elements or trace elements, the remainder are called macro-elements.

Early studies of element essentiality were carried out by J Sachs and W Knop in the 19th century and they considered that a mixture of KNO_3, $Ca(NO_3)_2$, KH_2PO_4, $MgSO_4$ and a salt of iron could supply all the elements needed. In fact, it was many years before we were able to put together the list shown in Table 4.1.

SAQ 4.2

Give a reason why it took so long to finalise a list of essential elements.

Further to the point made in SAQ 4.2, the requirement for boron can often be met by its release from borosilicate glassware, so essentiality studies should be carried out in pyrex glass vessels.

why minerals are essential

Let us now examine the functions of the elements to explain why they are essential and try to account for the concentration in which they are needed.

4.4 The function of minerals

Plants are composed of organic compounds and are made up of, primarily, carbon, oxygen and hydrogen. The elements are derived from carbon dioxide and water and are incorporated into cell constituents via photosynthesis. Since we discussed these processes thoroughly in Chapter 2, we will restrict ourselves to discussion of the function of mineral elements.

Some mineral elements are constituents of organic compounds which are present in considerable concentration in plants. For this reason the element itself is required in high amounts. This applies to nitrogen, phosphorus, sulphur, magnesium, and calcium.

nitrogen

Nitrogen occurs in all amino acids, proteins and nucleic acids and it is not surprising that more nitrogen is required than any other element except, hydrogen, oxygen and carbon.

phosphorus

Phosphorus is a constituent of phospholipids, which are important components of membranes. This is almost certainly where most phosphorus is used, although large quantities are found in nucleic acids. Many constituents of metabolism are present as phosphorylated intermediates, such as sugar phosphates, and each ATP molecule has three phosphate residues. These compounds, however, are usually present in relatively low concentration and they contribute in only a small way to the total phosphorus requirement.

sulphur

Sulphur is present in the amino acids cystine, cysteine and methionine and, therefore, in all proteins. It is also found in the tripeptide glutathione and these account for the bulk of the sulphur requirement. Sulphur is a part of numerous coenzymes, such as coenzyme A and ferredoxin, but these are present at relatively low concentration.

magnesium

One magnesium ion is present in the porphyrin ring of chlorophyll and so considerable quantities of magnesium are required in plant nutrition. It is also required as a cofactor for many enzymes, especially those involving the transfer of phosphate residues. Large amounts of magnesium ions are required to hold the nucleic acid and acid proteins in ribosomes together in the correct configuration. This, together with the metabolic roles of this element means that plants require substantial amounts of magnesium.

calcium

As we saw in Chapter 1, calcium occurs as calcium pectate in the middle lamella of cell walls and also acts as a cofactor for many enzymes especially those involving ATP hydrolysis. It is also thought to be involved in environmental signal transduction in plants.

The micro-nutrients, in the main, do not form part of organic compounds found in high concentration.

ion

Iron is a constituent of cytochromes and each haem group has one atom of iron in it. Iron is also a constituent of ferredoxin. However, cytochromes and ferredoxin are present in enzyme-like concentrations and iron is required in only micro-nutrient quantities.

Copper, zinc and molybdenum are, individually, found as cofactors of one or more enzymes and molybdenum is required as a cofactor for numerous enzymes. Copper is also found in plastocyanin, an electron carrier in the thylakoid. None of these compounds, however, are present in high concentration and the elements are required in only trace amounts.

SAQ 4.3

Hydrogen, oxygen and carbon excepted, two of the elements have been omitted from the above discussion. (You may need to refer to Table 4.1) Identify the two missing elements.

The two elements we missed out from our discussion are a little puzzling, especially boron. Boron does not form part of any essential organic compound and there are no enzymes which need it as a cofactor. Boron can form complexes with carbohydrates but there is no evidence that this occurs in plant tissues or that it would fulfil a specific function if it did. Boron, of course, is required in only trace amounts but, at the moment, we do not know why.

roles of potassium

Potassium is a macro-element and is next to nitrogen in terms of the amounts found in plant tissues. It is, therefore, surprising that it does not form part of any organic compound. It is required as a cofactor for over 40 enzymes but this does not seem able to explain the magnitude of its requirement. It is interesting to note, in this context, that the solution which can be collected as an exudate from phloem contains potassium at a concentration of about $77 \, mmol \, l^{-1}$ and that guard cells have even more, often more than $400 \, mmol \, l^{-1}$. Because of the different volumes occupied by guard cells as compared with the phloem the latter probably contains a higher proportion of the total plant potassium content. The function of potassium in guard cells, as you know, is to cause changes in guard cell π values which are instrumental in causing opening or closing of the stomata. Potassium is essential to this function because of the specificity of the potassium carrier in the guard cell plasmalemma. We cannot account for the level of potassium found in phloem exudate. It is present here at 10-100 times the concentration of any of the other elements. Potassium ions are, however, also required to maintain the structural integrity of ribosomes and this may, in part, account for the high intracellular concentration of these ions.

4.5 Lack of essential elements causes deficiency symptoms

deficiency symptoms

If all of the essential elements are available in a soil we would expect plants to grow very well. A deficiency of an element (by which we mean their presence is below optimal quantity) will result in reduced plant growth, often accompanied by a characteristic deficiency symptom. It is beyond the scope of this text to go into details here about all of the deficiency symptoms but it will be worthwhile to make a few points.

Perhaps the most characteristic symptom is that of phosphorus deficiency, in which the underside of the leaves becomes purple, especially on the veins. The deficiency of no other element causes the purple colour to develop. It is not known at present how phosphorus deficiency causes this pigmentation.

SAQ 4.4

What do you think the characteristics of magnesium deficiency might be?

We have just seen that a deficiency of magnesium causes leaf chlorosis but this symptom is not as diagnostic as the purple coloration of phosphorus deficiency. This is because, on the one hand, each porphyrin group has four nitrogen atoms and on the other, the element iron functions in chlorophyll synthesis. Thus a deficiency of nitrogen or iron will also cause chlorosis. Further, iron availability in the soil depends upon soil pH so chlorosis could develop even in the presence of apparently adequate amounts of iron. This effect of soil pH on mineral availability is not restricted to Fe. As shown in Figure 4.2, pH affects the availability of a wide range of minerals.

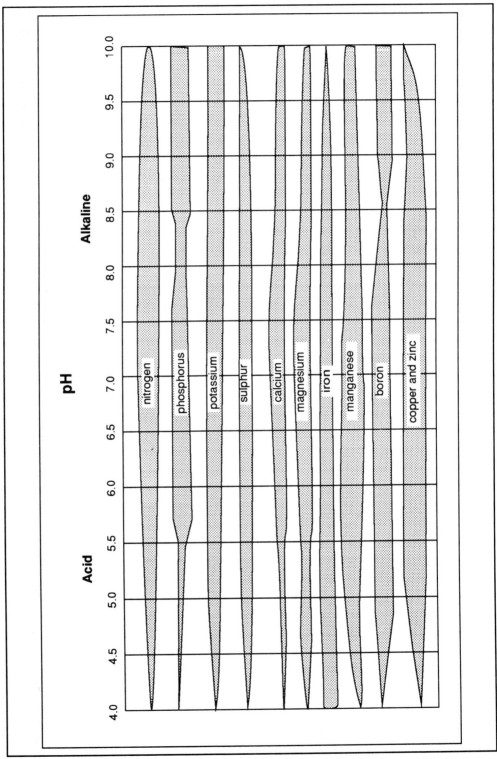

Figure 4.2 Influence of pH on availability of nutrients in soils. The vertical height of a zone indicates the relative availability of the element.

soil pH controls
mineral
availability

Species vary in the way they respond to soil pH. Plants such as azaleas and rhododendrons are acid-loving and grow best at pH 4.5, whereas the cabbage family, clover and alfalfa prefer pH 7.0. Most other plants perform well at pH values between these. The extreme cases of pH preference are not fully understood but it is likely to involve not only the effect of pH on mineral availability in the soil but also the effect of pH on the uptake mechanisms themselves. Despite the complexity of some deficiency symptoms they are still the first sign of a soil problem and would normally be supported by soil analysis and remedied by application of mixtures of elements.

4.6 Occurrence of the elements in soil

Table 4.1 shows that the elements phosphorus, sulphur, nitrogen, boron and molybdenum are available in the soil as phosphate, sulphate, nitrate, borate and molybdate anions, respectively. The other mineral elements are present as their cations. The elements originate from rocks of the lithosphere by the process of weathering, whereby complex crystalline structures are slowly broken down by physical and chemical processes to form soluble compounds. They become distributed in soils as a result of rain action but many are leached out and find their way via streams and rivers into lakes and oceans. Some are incorporated into living organisms and these are recycled when the organisms die and are degraded by bacteria and fungi. Nitrogen does not occur in rocks but does occur as natural deposits of nitrate, an important source being Chile saltpetre (KNO_3). These deposits are thought to originate by the decomposition of organisms many millions of years ago.

nitrogen
fixation

Nitrogen occurs in the soil predominantly as NO_3^-, NO_2^- and NH_4^+ ions, and these are obtained mainly by degradation of the nitrogenous compounds of dead plants and animals. Plants and animals cannot, themselves, make NO_3^-, NO_2^- or NH_4^+ from nitrogen but are dependent on the fixation of atmospheric nitrogen which occurs in two ways. Lightning causes the formation of hydroxyl free radicals, free oxygen and free hydrogen atoms from water and these attack atmospheric nitrogen to form nitric acid. This is washed to the ground by rain and accounts for the production of an estimated 2 kg of nitrate ha^{-1} yr^{-1}. The remaining fixation of nitrogen is brought about by the action of numerous micro-organisms and is called biological nitrogen fixation. This process, which produces approximately 300 kg of nitrate ha^{-1} yr^{-1}, will be described in detail later.

Soil is a very heterogeneous material containing a solid phase, a liquid phase and a gaseous phase. As a result of the processes described above the liquid phase contains numerous mineral ions in solution. Other mineral ions, predominantly cations, are adsorbed onto the surface of the soil particles but these are not firmly attached and are available to the plant as a result of cation exchange. This is the milieu in which plant roots find themselves, so let us now examine the absorption and transport of these minerals by plants.

4.7 The Casparian strip limits mineral diffusion in roots

Casparian strip

The Casparian strip refers to a zone of the walls of the endodermis which is impregnated with suberin, a waxy material (Figure 4.3).

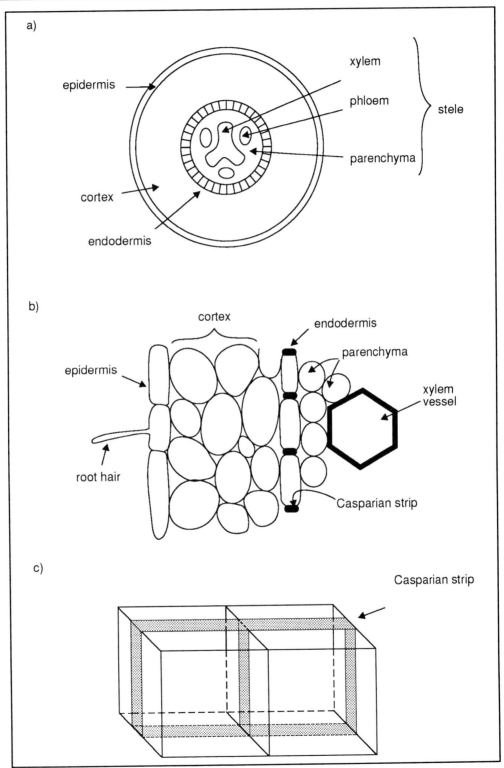

Figure 4.3 a) and b) cross sections of dicotyledonous root; c) diagram of two endodermal cells showing the Casparian strip.

The Casparian strip occurs on the four radial walls of the endodermis but not on the two tangential walls. The waxy nature of suberin means that the Casparian strip is waterproof. Further, the plasmalemma of the endodermal cell is firmly attached to the cell wall in the zone of the strip.

SAQ 4.5

Figure 4.3 b) shows a cross section of a root with a root hair protruding into the soil. What part of the root hair is in direct contact with the soil? Is it the apoplast or the symplast?

Since the apoplast is in direct contact with the soil, if there is a suitable chemical potential gradient, minerals will be able to diffuse from the soil into the solution present in the root apoplast. However, minerals will not be able to travel past the endodermis because of the waterproofing nature of the suberin. Further, minerals will not be able to sneak through between the wall and the plasmalemma because these two are joined together. Thus, if minerals are to reach the stele, they must enter the symplasm, because the symplasm of the cortex is in continuity with that of the endodermis and stelar parenchyma.

ion absorption
The actual position at which ions cross the plasmalemma to enter the symplasm varies but is usually in the root hairs, the epidermis or outer cortical cells. The important point here is not where the ion crosses the plasmalemma but the fact that it passes through at all. As you remember, the plasmalemma is a selectively permeable membrane and by this property the plant controls what enters the symplasm. The absorption of ions is a carrier-mediated process and, since the ions are accumulated against their concentration gradient, the process requires metabolic energy. The selectivity is also very important here. The mineral ion regime of the root cortex, ie the mineral content and their concentration, can be different from that of the soil solution and this is because of the selective absorption of ions by the plasmalemma. Detailed studies of ion absorption can be carried out most easily using algae such as *Hydrodictyon*. These have such large cells that ion-selective microelectrodes can be inserted directly into them.

∏ Examine the data of Table 4.2, compare the concentrations of the ions listed in the medium and in the cytoplasm and make comments about them.

	Medium (mmol l^{-1})	Cytoplasm (mmol l^{-1})
K$^+$	0.1	93
Na$^+$	1.0	51
Cl$^-$	1.3	58

Table 4.2 Values for the concentrations of certain ions in cells of *Hydodictyon* and in a typical medium.

All three ions are present in the cytoplasm at concentrations well above those of the medium, implying absorption by an energy-requiring process. Note that K^+ concentration shows a 930 fold increase, which is much larger than the increase for Cl^- and Na^+. The situation regarding Na^+ is interesting. Although it is not a required element many plants, particularly aquatic ones, accumulate it when grown in natural waters.

4.8 How do ions enter the xylem vessels and tracheids?

For many years it was considered that stelar parenchyma cells were 'leaky' and ions simply diffused out into the xylem vessels and tracheids. Remember that these latter two types of cell are dead and are, therefore, part of the apoplasm. Thus, entry into the xylem was considered to occur by diffusion through the plasmalemma of the parenchyma. Evidence more recently accumulated does not agree with this. Thus, the determination of *in situ* ionic concentrations in the stele by several different methods showed them to be higher than in the cortex. Further, the determination of the concentration of ions in the exudate from severed stems showed it to be higher than that in the root cortex. This would imply that energy was used to transport the ions out, (active absorption in reverse). This process is called secretion. It is now generally accepted that the stelar parenchyma cells are not 'leaky' but actively transport ions into the xylem.

SAQ 4.6	Experiments show that the mineral ion content of shoots is very often different from that of roots. You should be able to suggest one phenomenon to account for this but try to think of a second. If you cannot get the second one look at the hint in the response.

In addition to aquatic plants, many terrestrial plants including sugar beet, absorb sodium from the soil. Sodium seems to be toxic to the shoots of terrestrial plants but the problem is avoided by retaining sodium ions in their roots, roots being much less susceptible than shoots.

SAQ 4.7	We have seen that ions in the xylem are now back in the apoplast. They are present there at higher concentration than in the cortex and in the soil. Why do they not simply diffuse out through the apoplasm?

The final point to make here is that it is not only potentially toxic ions which are held back in the roots. Root cells need minerals for their own purposes and usually transport to the shoot no more than 30% of that which they absorb. The most active zone in absorption is the elongation zone in the root tip and the zone most active in translocation of minerals is the zone which has just completed cellular differentiation. In barley this is typically 30mm behind the tip (Figure 4.4).

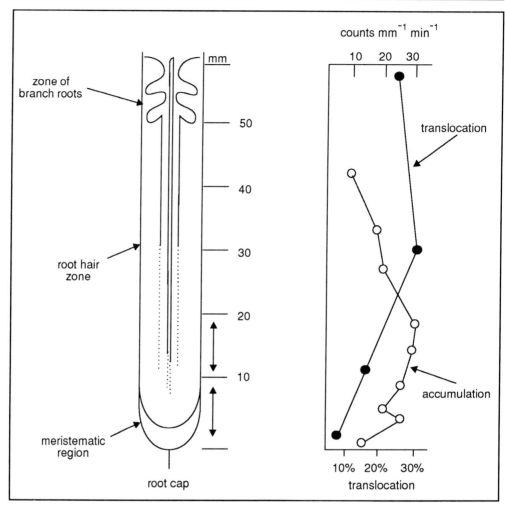

Figure 4.4 Absorption and translocation of minerals by barley roots. Translocation in particular zones is expressed as a percentage of the total. Accumulation is shown as counts min^{-1} of radiolabelled minerals absorbed mm^{-1} of root.

4.9 Minerals are given a free ride to the shoot

mineral
transport

Once minerals are deposited in the xylem, they find themselves in a moving stream of fluid, the transpiration stream, and are carried along in it. The transpiration stream delivers minerals to all parts of the shoot which are transpiring. This includes all parts except young leaves and developing seeds and fruits. These exceptions will be examined later when we discuss the function of the phloem.

The transpiration stream delivers minerals to the leaf apoplasm. Thus, the cells of the leaf are bathed in a solution of mineral salts and it is from this solution that cells remove the elements they need. Although detailed studies of mineral ion absorption by leaf cells are few, it is known that carriers are present which transfer specific elements into cells and that cells possess carriers only for those elements which they need. Do not forget that there are many different cell types in a leaf and they do not have identical mineral requirements. One example will suffice. Epidermal cells do not have chloroplasts and do not produce chlorophyll so they have a much lower magnesium requirement than mesophyll cells. It was mentioned earlier that the guard cells are not connected by plasmodesmata to subsidiary cells of the epidermis. This allows the guard cells to maintain ionic regimes quite different from the epidermal cells. Although it is difficult to prove unequivocally, it is suspected that there are no plasmodesmata linking the mesophyll cells with the epidermal cells of leaves.

SAQ 4.8

See if you can now remember the three places in which mineral selection can occur during the transport of minerals from the soil to the leaf. What are some of the consequences of these selections?

4.10 Plants can transport ions upward while transpiration is not occurring

When the roots are exposed to moist warm soil containing an adequate supply of nutrients, and the leaves are exposed to humid air, water will enter the roots and eventually be exuded from the leaves. The exudation occurs through hydathodes, which are modified stomata and the process is referred to as guttation (Figure 4.5).

guttation

Figure 4.5 Guttation.

This liquid is most commonly seen as droplets clinging to leaves early in the morning and analysis of its constituents show that salts are present at a higher concentration than in the soil solution. This is produced in the following way. Salts are actively absorbed by the root and secreted into the xylem but transpiration is not occurring because of the higher humidity of the surrounding air. The roots, however, continue to transport salts, under the influence of the warm temperature. The high salt content of the xylem reduces its water potential and water moves in from the stelar symplasm by osmosis. This generates a positive pressure which forces the fluid upwards, into the shoot and leaf apoplasm and out through the hydathodes. This pressure is root pressure described in the previous chapter.

4.11 Some plants excrete salt

salt excretion

Plants which inhabit coastal zones and saline deserts are in the physiologically difficult position of having their roots exposed to solutions of high salt content. Plants which live on the banks of estuaries face periodic fluctuations in the content of their soil solution. These plants balance their internal π values with those in their environment by actively accumulating sodium ions, electrical neutrality being maintained by passive chloride absorption. In these plants, sodium is transported to the shoot where the overall salt balance of the plant is maintained by the action of salt glands. These are specialised groups of cells on the leaf surface which actively transport Na^+ and Cl^- out of the cells onto the leaf surface, from which they are removed by the action of wind and rain (Figure 4.6).

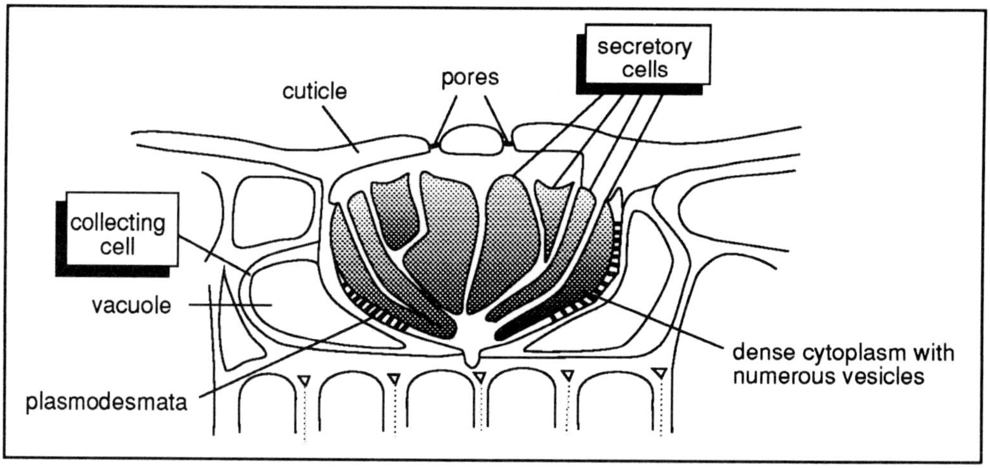

Figure 4.6 A salt gland of sea lavender (*Limonium latifolium*). (Redrawn from Plant Physiology, OU, Milton Keynes, UK, Figure 3.30).

The most striking mechanism for salt removal is shown by those plants such as desert salt bushes, which have salt hairs. These consist of a stalk and a bladder cell (Figure 4.7). Salt is accumulated in the bladder cell which eventually bursts, releasing the salt.

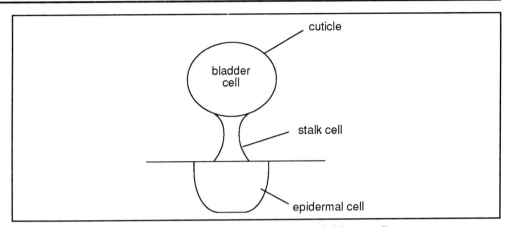

Figure 4.7 Diagram of stalk and bladder cell found in the salt bush, *Atriplex spondiosa*.

Sodium is an essential element for these plants and its absorption allows the plant to obtain water from its environment despite the latter's very high mineral concentration.

SAQ 4.9

The salt marsh plant, *Spartina maritima*, is an example of a plant which secretes salt as described above. Explain how you would demonstrate that sodium is essential for this species.

SAQ 4.10

Data relating to the concentration of three cations in the external solution, the exudate obtained from xylem and the gluttation fluid of marrow seedlings are shown in Table 4.3.

What general conclusions can be drawn about the processes of absorption and secretion of cations by marrow seedlings.

Ion	External concentration (mmol l^{-1})	Concentration in xylem exudate (mmol l^{-1})	Concentration in guttation fluid (mmol l^{-1})
K^+	0.5	2.8	0.5
Na^+	0.25	0.3	0.7
Ca^{2+}	0.15	5.1	0.9

Table 4.3 Changes in the concentration of ions during their passage through the xylem and leaves of marrow seedlings.

SAQ 4.11

Part of the bark of an intact young willow tree grown in hydroponics was carefully separated from the wood and waxed paper inserted between them, as shown below.

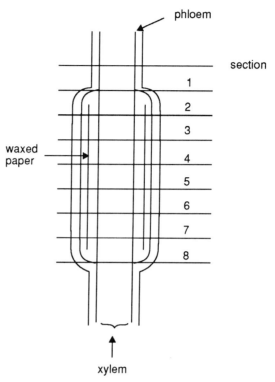

A radioactive potassium solution was added to the hydroponic solution and transpiration allowed to continue for five hours. The experimental portion of the stem was then divided into 8 sections and the radioactivity of the xylem and phloem of each section determined. The concentration of potassium was then calculated from this giving the results below.

What conclusions can be drawn from this experiment concerning the transport of potassium?

	Potassium concentration (ppm)			
	Stripped plant		**Unstripped plant**	
Section	Bark	Wood	Bark	Wood
1	53.0	47	64	56
2	11.6	119		
3	0.9	122		
4	0.7	112	87	69
5	<0.3	98		
6	<0.3	108		
7	20.0	113		
8	84.0	58	74	67

(Data from P Stout and D Hoagland, American Journal of Botany (1939) 26:3201).

4.12 Plants can absorb minerals through their leaves

We saw above that the leaves are supplied with minerals through the solution in the apoplasm. The mineral content of this solution can be quickly modified by applying the minerals directly to the leaves as a foliar spray. This is a very effective way of applying mineral fertilisers to crops. This topic is covered in more detail in the partner BIOTOL text, 'Crop Productivity'.

4.13 Nitrogen is a special case

Nitrogen is the major limiting mineral resource for most plant species and its acquisition and assimilation is second in importance only to that of carbon. The provision of high-quality protein-rich food for animals is very much dependent upon the availability of sufficient nitrogen for plant growth and development.

nitrate reduction

Nitrogen is absorbed primarily as soil nitrate and this is converted to ammonium ions prior to incorporation into amino acids. This occurs in both the root and the shoot. In the first step of this pathway nitrate is reduced to nitrite by nitrate reductase. The reaction occurs in the cytosol:

$$NO_3^- + NADH + H^+ \longrightarrow NO_2^- + NAD^+ + H_2O$$
$$\text{nitrate} \qquad\qquad\qquad \text{nitrite}$$

The next step involves the reduction of nitrite to ammonia, catalysed by nitrite reductase. This enzyme occurs in plastids. In roots and other non-green tissue the reaction is as follows:

$$NO_2^- + 3\,NADH + 5\,H^+ \longrightarrow NH_4^+ + 3\,NAD^+ + 2\,H_2O$$
$$\text{nitrite} \qquad\qquad\qquad \text{ammonium}$$
$$\text{ions}$$

In green tissues the reducing power comes from photosynthesis via ferredoxin:

$$NO_2^- + 6\,Fd\,(red) + 8\,H^+ \longrightarrow NH_4^+ + 6\,Fd\,(ox) + 2\,H_2O$$

Ammonia and ammonium ions do not accumulate in plant tissues. They are toxic above certain levels and plants utilise numerous mechanisms to avoid accumulation to these levels.

ammonia assimilation

The assimilation of ammonia nitrogen into organic compounds can occur in two ways and we will examine them in some detail to show how plant biochemists tackle the problem of deciding whether both or just one pathway operates.

Pathway 1

$$\alpha - \text{oxoglutarate} + NADH + NH_4^+ \longrightarrow \text{glutamate} + NAD^+ + H_2O$$

This is the reductive amination of α-oxoglutarate by glutamate dehydrogenase (GDH).

Pathway 2

$$\text{glutamate} + NH_4^+ + ATP \longrightarrow \text{glutamine} + ADP + Pi$$

This reaction is catalysed by the enzyme glutamine synthetase (GS). Glutamine can then be converted to glutamate:

$$\text{glutamine} + \alpha - \text{oxoglutarate} \longrightarrow 2 \text{ glutamate}$$

by the enzyme glutamine-oxoglutarate aminotranferase (GOGAT).

GS/GOGAT

The last enzyme in the sequence was discovered in 1970. Previous to that GDH was considered to be responsible for glutamate synthesis in plants. The current view, however, is that NH_4^+ assimilation and glutamate production is predominantly through the GS/GOGAT pathway. Let us examine some of the evidence.

- The Km (NH_4^+) for GDH (30-80mmol l^{-1}) is much higher than that for GS (1.0-5.0 mmol l^{-1}) so GDH would not be able to compete very well with GS for the available NH_4^+.

- The GS inhibitor methionine suphoxime and GOGAT inhibitor azaserine do not affect GDH activity but they prevent the incorporation of labelled nitrogen $^{15}N\text{-}NH_4^+$ into amino acids.

- In the absence of inhibitors, labelled nitrogen from $^{15}N\text{-}NO_3^-$ or $^{15}N\text{-}NH_4^+$ appears first in the amide group of glutamine and then in glutamate.

Thus, GDH does not appear to be involved with NH_4^+ incorporation into amino acids at normal concentrations of NH_4^+. Its Km value is within the range considered to be toxic for plant cells. Thus, its function may be to remove NH_4^+ when it is present at these high levels.

amino transferases

Once incorporated into glutamate, nitrogen can be incorporated into other amino acids by the process of transamination, catalysed by aminotransferases. An example is the formation of aspartate by aspartate aminotransferase.

$$\text{glutamate} + \text{oxaloacetate} \longrightarrow \text{aspartate} + \alpha - \text{oxoglutarate}$$

The principle is the transfer of the amino group of glutamate to an α-oxo acid to form the appropriate amino acid, α-oxoglutarate being made available to be re- utilised by GOGAT. Plants possess a full complement of aminotransferases which enable them to make the full complement of amino acids needed for protein synthesis. In addition, the enzyme asparagine synthetase catalyses the conversion of aspartic acid into its amide asparagine. This, too, can act as a donor of amino groups.

It was mentioned earlier that the reduction of nitrate to ammonium ions occurs both in roots and shoots. Under conditions of low availability of nitrate, all of the nitrate is reduced in the root and nitrogen is exported as glutamine, asparagine and amino acids. Under conditions of high nitrate availability, nitrate itself is exported to the shoot where its reduction and incorporation into amino acids occur.

SAQ 4.12

The diagram below shows steps in the incorporation of nitrogen into amino acids. Fill in the blank squares. Squares 1-5 are enzymes; Square 6 is an organic compound whose structure should be drawn.

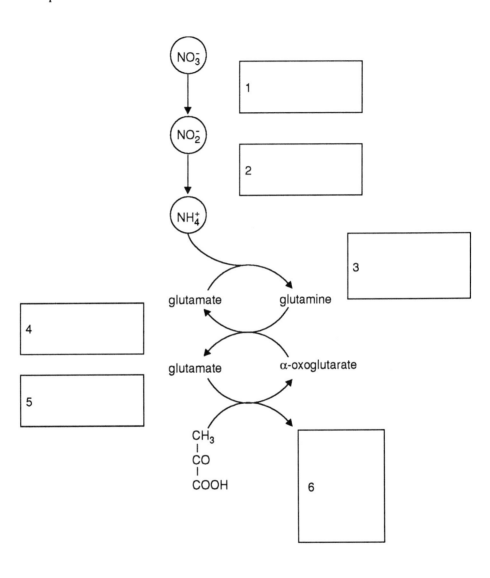

4.14 Nitrogen fixation

symbiotic
nitrogen
fixation

The fixation of gaseous nitrogen can be carried out by a number of micro-organisms. These include cyanobacteria (blue-green algae), free-living bacteria, such as members of the genera *Azotobacter*, *Klebsiella* and *Clostridium* and bacteria which are found predominantly in symbiotic association with plant roots, such as *Rhizobium*, which are found in legumes. The roots of Alder and Casuarina trees can fix nitrogen by virtue of their infection by the actinomycete, *Frankia*. Similarly the water fern *Azolla* is able to fix nitrogen as a result of its symbiotic association with the cyanobacterium *Anabaena*, which is capable of fixing as much as 1 kg ha^{-1} day^{-1} of atmospheric nitrogen in flooded rice fields. Thus, nitrogen fixing-ability has evolved independently in a wide range of micro-organisms, but in all cases it is catalysed by the enzyme nitrogenase, an enzyme which has never been found in eucaryotes. The conversion of nitrogen to NH_4^+ is very energy-demanding and is accompanied by the release of hydrogen gas:

$$N_2 + 12ATP + 6e^- + 8H^+ \longrightarrow NH_4^+ + 12ADP + 12Pi$$

$$4ATP + 2e^- + 2H^+ \longrightarrow H_2 + 4ADP + 4Pi$$

The whole reaction is catalysed by the enzyme nitrogenase, which adds one electron and one H$^+$ at a time to diatomic nitrogen, the N-N bond being broken in the latter stage of the process. Nitrogenase is very sensitive to oxygen which irreversibly inhibits the enzyme. In anaerobic organisms, such as *Clostridium*, O_2 does not present a problem but it does in aerobic organisms and numerous strategies have evolved to prevent O_2 reaching the enzyme. In *Azotobacter* this is achieved by high levels of respiration, whilst in the photosynthetic cyanobacterium *Gloeotheca* nitrogen fixation occurs only at night. A rather unusual method is used by *Rhizobium*. The rhizobia and their hosts jointly produce haem-containing molecules called leghaemoglobins. The host plant cell produce the apoprotein while the bacteria appear to be responsible for producing the haem moieties. This behaves similarly to haemoglobin in our blood and binds oxygen except that it binds oxygen rather more strongly. Thus it reduces the amount of oxygen available to the *Rhizobuim*. In other words, it creates a micro-aerophilic environment, allowing the *Rhizobuim* sufficient oxygen to carry out respiration, but keeps the oxygen concentration so low as to avoid inhibition of the rhizobial nitrogenase.

protection against damage by oxygen

Most of the fixed nitrogen is initially utilised by the organisms which make it but much of it finds its way into the soil. Ammonium ions are very soluble and can easily be leached away but much of it is taken up by soil organisms before this can happen. Two are or particular importance as far as plants are concerned. These are bacteria of the genera *Nitrosomonas* and *Nitrobacter* which carry out the following reactions:

$$NH_4^+ \xrightarrow{\text{nitrosomonas}} NO_2^- \xrightarrow{\text{nitrobacter}} NO_3^-$$

Nitrosomonas oxidises NH_4^+ to NO_2^- and releases it to the soil. *Nitrobacter* absorbs NO_2^-, oxidises it to NO_3^- and releases it. Thus, the production of nitrate from gaseous nitrogen involves the operation of three groups of organisms. The term nitrification is given to the process of converting NH_4^+ to NO_3^-.

4.15 Symbiotic nitrogen fixation involves the production of root nodules

Symbiotic nitrogen fixation occurs in nodules which are formed on the root in response to infection by the bacterium (Figure 4.8).

a)

b)

Figure 4.8 Root nodules on a) soybean and b) the alder tree.

The infection process and formation of the nodule has been studied most closely in the case of legumes which are infected by *Rhizobium*. Here it appears that exudates from the root first attract the bacteria to the root surface in the region of root hairs. Secretions from the bacteria then cause a remarkable train of events, as shown in Figure 4.9.

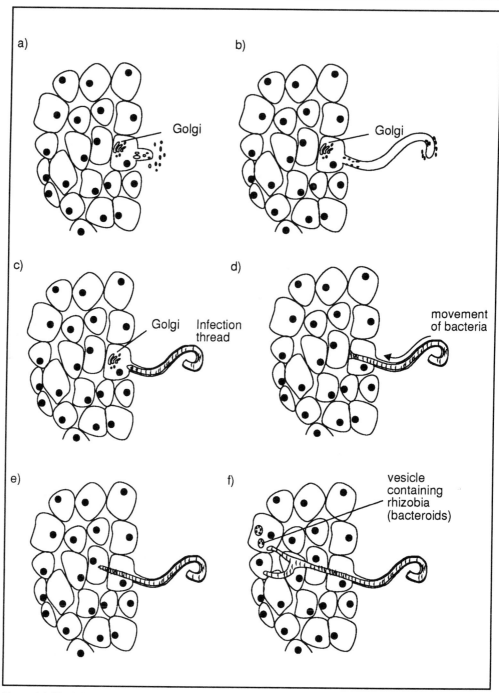

Figure 4.9 The infection process. a) and b) Bacteria bind to emerging root hair which exhibits unusual curling growth. A membranous tubule is formed by the Golgi apparatus. c) and d). The tubule extends into the body of the cell and fuses with the cell plasmalemma. At the curled end it fuses with the membrane and degrades the wall, the bacteria migrating along the tubule which is now referred to as the infection thread. e) and f) Bacteria are released into the apoplast and cause the formation of a secondary infection thread which ramifies deeply into the cortex. Finally, vesicles are pinched off consisting of bacteria surrounded by plasmalemma, termed bacteroids.

The vesicles containing bacteria release cell division factors into the plant cytosol. This causes localised cell division resulting in a swelling which becomes the nodule. The mature nodule is 4-5mm in diameter and contain hundreds of cells containing vesicles with bacteria inside. The nodules produce leghaemoglobin which, as we noted above, binds O_2 and prevents inhibition of nitrogenase activity. A measure of the complexity of this situation is shown by the fact that the bacteria produce the haem portion and the plant cells the globin portion of the leghaemoglobin molecule. As a result of the presence of leghaemoglobin the inside of the nodules is a delicate shade of pink.

symbiosis

The bacteria fix nitrogen which has diffused into the plant from the atmosphere and produce NH_4^+, in the manner described above. Considerable quantities of NH_4^+ are released from the bacteria into the symplasm inside the nodule and is acted upon in exactly the same way as NH_4^+ produced from nitrate. Thus, the plant benefits from the association. The bacteria absorb carbohydrate substrates from the plant symplasm and so the bacteria also benefit. There are numerous examples in biology of two organisms living together. The term symbiosis is applied to the relationship when, as here, both organisms benefit. Because of the nitrogen fixation occurring in the nodules, these plants are much less dependent upon the application of nitrogen-containing fertiliser. Apart from the cost benefit, a reduction in nitrate application might result in less nitrate run off to drinking water supplies. The process of nitrogen fixation has received close attention by biochemists and, more recently, molecular biologists. These studies have revealed species specificity between bacterium and host and the range of methods by which infection is achieved. The long-term hope of these studies is that species which do not fix nitrogen could be made to do so thus reducing the need for fertiliser applications. Some success has been achieved. The genes controlling nitrogen fixation have been isolated and it is possible, for example, to cause the formation of nodules on the roots of rice. Unfortunately, the nodules do not fix nitrogen so there is still much to be learned about nitrogen fixation.

4.16 Mycorrhizal fungi aid mineral ion absorption

ectotrophic and vesicular - arbuscular mycorrhizae

Mycorrhizae are associations between certain fungi and plant roots which increase the ability of the root to obtain minerals from the soil. Most of the vegetation of the world is infected with mycorrhizal fungi and, because both fungus and host benefit from the association this is another example of symbiosis. The infection is limited to the roots and two major classes of mycorrhizal fungi are involved, termed the ectotrophic mycorrhizae (ECM) and vesicular-arbuscular mycorrhizae (VAM). The ECM form a 40μm thick sheath of mycelium around the roots, the mycelium also penetrating inwards between the cells of the cortex and outwards between the soil particles (Figure 4.10a). ECM occur exclusively on the roots of trees and forms what is called a Hartig net.

VAM are less obvious on external inspection because they do not produce a sheath, simply consisting of internal and external mycelium. The internal mycelium forms highly branched structures within single cells, called arbuscles, as well as intercellular vesicles and hyphae (Figure 4.10b).

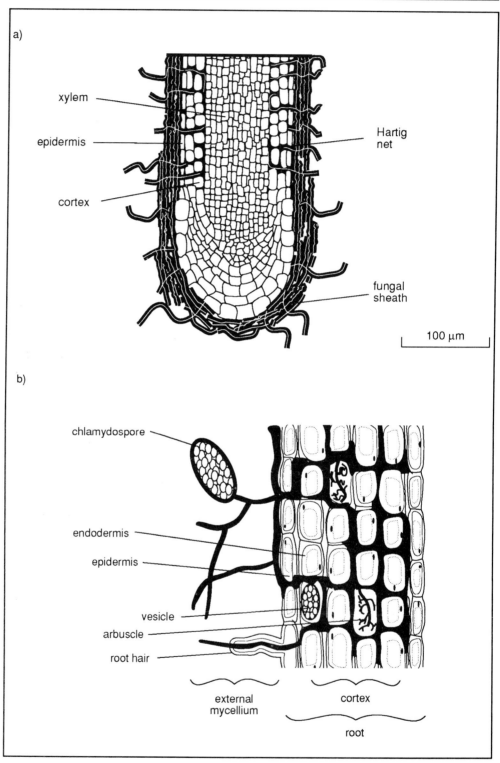

Figure 4.10 Diagram of a root infected with a) ectotrophic mycorrhizal fungi and b) with vesicular -
arbuscular mycorrhizal fungi.

A number of ECM fungi can be grown in pure culture so knowledge about ECM fungi was developed much earlier than for VAM, which are obligate symbionts and, therefore, much more difficult to culture. It has been known since the 1930s that ECM increase the yield of forest trees through an improved absorption of minerals particularly phosphate, and the artificial inoculation of forest nurseries with appropriate fungi is now an accepted practice. It was not until 1960 that evidence was obtained for an effect of VAM on mineral absorption by agricultural crops. These results showed that whereas VAM plants could grow very well on phosphorus-deficient soils, non-infected plants were very stunted. Subsequent work has not only confirmed the results with phosphorus but also demonstrated an improved ability to obtain micro-nutrients such as copper and zinc. The implication of the above is that the phosphate and micro-nutrients absorbed by the fungal mycelium are transferred to the roots of the host plants. Direct analysis has confirmed that this does, in fact, take place. What is unclear, however, is how the minerals move from fungus to root cell. Some evidence suggests that the minerals simply diffuse out of the fungus but other studies suggest that nutrients are released only when the fungal cells die and disintegrate. Interestingly, a key factor governing the infection of a plant cell by VAM is the phosphorus status of the plant. Phosphorus deficiency promotes infection, whereas growth in well fertilised soil suppresses infection. Whereas ECM inocula are available for use in tree nurseries, a cheap source of VAM inoculum is not yet available thus limiting the extent of its commercial use.

benefits from the mycorrhizal condition

SAQ 4.13

Pine seedings were grown with and without ectomycorrhizal infection and the following results obtained.

	Growth (g dry weight per plant)	Content of N,P,K: (mg per dry weight)		
		N	P	K
no mycorrhiza	0.26	9.20	0.85	8.70
with mycorrhiza	2.81	13.80	2.0	12.67

Calculate the proportional increase in nutrient absorption and, based on your knowledge of the action of mycorrhizae on P absorption, try to explain the N and K results.

4.17 Carnivorous plants

Six families of flowering plants have evolved complex and intricate structures which allow them to trap, digest and absorb insects. The plants concerned, examples of which are shown in Figure 4.11, are all green photosynthetic plants so it was puzzling to decide what they might gain from the carnivorous habit.

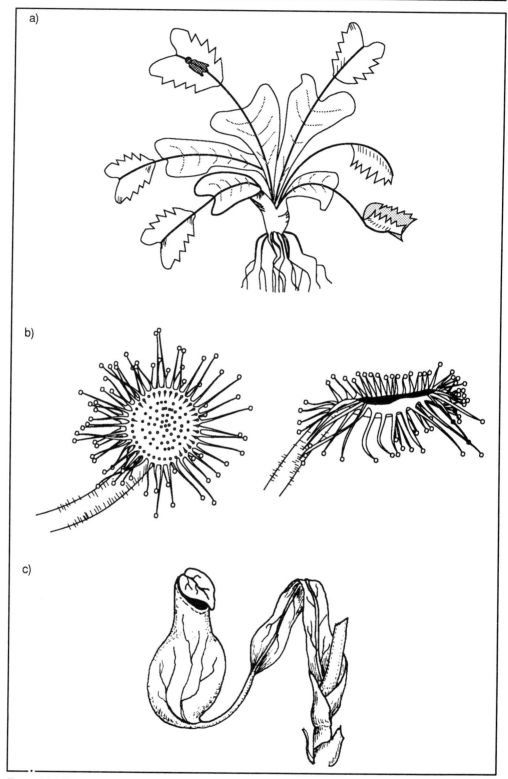

Figure 4-11 Examples of carnivoorous plants. a) Venus fly trap (*Dionala muscipula*), b) Sundew (*Drosera*) and c) Pitcher plant (*Nepenthes*).

Recent experiments have shown that these plants show enhanced growth as a result of trapping and digesting insects, even when well supplied with nutrients through their roots. The elements N, S and P seem to be particularly important in this context and it might suggest a slight deficiency in the plants ability to absorb these minerals through their roots.

4.18 How do plants survive an overdose of metal?

heavy metal
toxicity

We have seen above that plants need metals if they are to grow. Some metals are required in large amounts, such as potassium and calcium and an excess of either is usually not harmful. Metals such as copper and zinc are required in much lower amounts, predominantly for the activity of certain enzymes. However, exposure of these enzymes to anything more than a trace of copper or zinc causes irreversible damage. Soils which contain high levels of copper or zinc or a range of other metals are usually toxic. Soils normally used for agricultural purposes do not contain toxic levels of minerals, but considerable areas of land do contain toxic levels and so cannot be used for agriculture. These areas either contain the high levels naturally or artificially due to the spreading of industrial wastes. However, soils which contain high natural levels of minerals, such as those over ore veins, are not totally barren, they often support a specialised flora made up of species which have evolved some degree of tolerance to metals. These plants are usually found to be the first and, sometimes, the only colonisers on spoil tips.

exclusion
mechanisms

Studies of these plants reveal two methods for resisting the effects of metals. In some species the metal element is prevented from entering the plant. In our discussion above (Section 4.6) we noted that mineral ions diffuse into the root apoplasm from where they are selectively absorbed into the symplasm. Plants which can largely prevent the absorption of the toxic element can survive in its presence. Such mechanisms are called exclusion mechanisms.

phytochelatins

A second mechanism is to convert the mineral into something which is less harmful. This is a tolerance mechanism and very important strides have been made recently in understanding them. These species produce large quantities of specific proteins which bind metal elements and effectively take them out of solution. Thus, they can do no harm to enzymes in the root and they are not available for export to the shoot. These proteins are called phytochelatins and they have been found in numerous metal-tolerant species. These metal-tolerant species are not often themselves important as crop plants but molecular biologists are now studying the production of phytochelatins in the hope that it will be possible to engineer crop plants to produce them. This would open up for agricultural use large areas of land which are presently capable of supporting only limited plant growth.

SAQ 4.14

Experiments with a range of strains of *Agrostis tenuis* revealed the results stylised in the figure below.

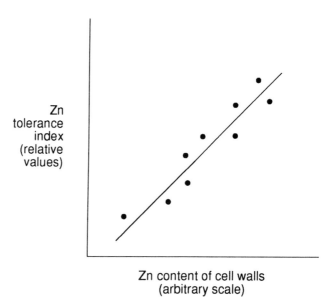

(After Turner and Marshall, 1972; New Phytologist 71, 671-676).

Each point on the graph indicates a separate strain which was tested for its ability to tolerate zinc and for Zn content of its cell walls. The line was drawn by regression analysis.

What explanation can you give for these results?

4.19 Young growing tissues obtain minerals through the phloem

We noted in section 4.8 that the transpiration stream will, by definition, deliver ions only to transpiring tissues. Young tissues at the shoot apex have a need for minerals before they actually start to transpire. Thus, transpiration, in this case, is not responsible for delivery to these young tissues. Minerals are delivered to them not through the xylem but through the phloem and this is the topic of the next chapter.

Summary and objectives

This chapter described the development of our knowledge of mineral nutrition, how hydroponics was used to decide which minerals were essential for plant growth and the functions these minerals perform in the plant. The origin of minerals in soils was discussed and the role of the Casparian strip in absorption described. An account was given of the distribution of ions in the transpiration stream and of the strategies shown by plants which live in estuaries and salt marshes. The role of glutamine synthetase and GOGAT in the assimilation of nitrogen was described. The importance of nitrogen fixation and the structure and function of root nodules explained. The occurrence of mycorrhizal associations and their effect on mineral absorption were also described and the chapter finished with a discussion of the mechanism which allow certain species to survive on land which contains abnormally high concentrations of minerals. Now that you have completed this chapter, you should be able to;

- demonstrate an understanding of how hydroponics was used to decide which elements are essential;

- list the essential minerals, describe their functions and show an understanding of the meaning of the term deficiency symptoms;

- describe the origin of minerals in soil, explain how they can be selectively absorbed and transported from the root to the shoot and into the leaf cells and interpret experimental data relevant to this;

- explain the pathway of NO_3^- assimilation and the role of free-living and symbiotic nitrogen fixing organisms in the nitrogen economy of plants;

- describe the structure of mycorrhizae and propose an explanation for their effect on the mineral status of roots;

- describe the two basic mechanisms that enable certain plants to cope with high concentrations of toxic elements;

- interpret data relating to the uptake and transport of mineral elements in plants.

Transport of organic compounds

Transport of organic compounds

5.1 Introduction

photosynthate

We noted in the previous chapter that the division of labour between roots and shoots of terrestrial plants resulted in the need for a system to transport minerals from the roots to the shoots. The same division of labour results in the need for a system to transport the products of photosynthesis (ie photosynthate) from the leaves to the roots, which do not photosynthesise.Transport of photosythate is also required to areas such as stems and young leaves which are not self-sufficient in this context.This transport system is the topic of this chapter. We will study its structure and function, describe efforts to elucidate the manner of its control and give an outline of attempts to manipulate the process for the benefit of mankind.

5.2 Organic compounds are transported in the phloem

Although it was suggested in 1837 by Hartig that sugars are transported in the phloem, it was only as recently as 1956 that Biddulph provided definitive evidence of this process. Here we will describe some of the early work as it is a suitable way to introduce some of the methods used in studying phloem function.

girdling

In early experiments, girdling of a woody shoot by removal of a ring of bark, left the xylem intact and resulted in the swelling of the tissues above the girdle and the shrinking of those below. Analysis of the swollen tissues showed them to contain high concentrations of sucrose, but this was present in the xylem as well as the phloem. It was not established at that time that the xylem functioned in upward transport of water and minerals, so the question of the site of sucrose transport remained unproved.

Scientists then moved on to use various surgical treatments, in which strips of phloem were teased apart and kept separate by application of Vaseline (petroleum jelly) or by the insertion of waxed paper. The results of one such experiment are described in Figure 5.1.

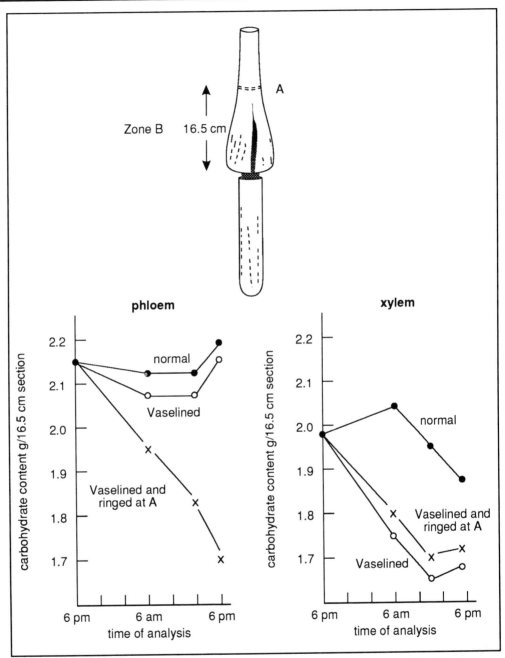

Figure 5.1 Effects of separating the xylem and phloem by a layer of Vaseline on the translocation of carbohydrates into zone B. Normal = section of phloem lifted and replaced without Vaseline; Vaselined = phloem lifted but replaced after applying Vaseline; Vaselined and ringed at A = the same as Vaselined but the stem girdled at A (data of Mason and Maskell 1928, Annals of Botany *42*, 190-253).

In the 'normal' plants the levels of carbohydrate were high in both the phloem and xylem in zone B but the application of Vaseline caused a marked reduction in the level in the xylem. Removal of a ring of phloem by girdling the stem severely reduced the level of carbohydrate in the phloem and this experiment strongly indicated that the phloem was the normal channel for the movement of carbohydrate. It did, however, also provide evidence for the lateral movement of carbohydrate from phloem to xylem. Definitive evidence for the phloem as the site of transport was obtained when radioisotopes became available. The exposure of leaves to $^{14}CO_2$ results in the production of radio-labelled sucrose as the principal form of mobile photosynthate. If sections of stems are cut from such plants, the distribution of labelled sucrose can be determined by autoradiography. In this technique, a film of photographic emulsion is applied to the section and the two incubated together for several days. Radio-isotopes cause a reaction in the film which is similar to that caused by light exposure. Development of the film shows black dots in positions where it has been in contact with radio-isotopes. Careful alignment of the film with the section revealed the presence of black exposure zones strongly concentrated over the phloem sieve cells. This showed not only that sugars are transported in the phloem but also that the particular cell involved was the sieve cell.

SAQ 5.1

The diagram below shows two plants in which the leaves are exposed to $^{14}CO_2$ and the roots to ^{32}P. In one plant, a jet of steam has been used to kill a narrow portion of the stem in a process referred to as steam ringing. Data for the distribution of label is shown as the percentage of total label recovered.

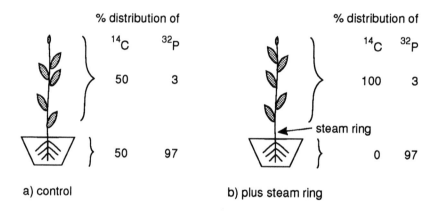

a) control   b) plus steam ring

The leaves in both plants remained turgid throughout the experiment. What interpretation would you put on these results. There is a hint in the response about the action of steam ringing, should you need it.

5.3 A solution moves in the phloem by mass flow

Theoretically, compounds could move in the phloem by diffusion but the application of Fick's Law shows that this is inconceivable. Fick's Law enables us to determine how long it would take a substance, at a given distance from the starting point, to reach half the concentration at the starting point We can denote this time as t_c.

Fick's Law can be represented as follows:

$$t_c = \frac{(\text{distance})^2 \times K}{D}$$

where K is a constant which depends on the geometry of the system and D is the diffusion constant of the substance.

$\prod$ If we use typical values of 1 for K and 10^{-9} m^2s^{-1} for D, we can calculate how long t_c will be if the distance is a) 50μm and b) 1m. Do this calculation and check your answers with those provided in the table below.

Distance to travel	Time to achieve half concentration
a) 50μm	2.5s
b) 1m	10^9s = 24yrs

Table 5.1 Typical time needed to achieve half concentrations at various distances from a source.

Think back to the sizes of cells given in Chapter 1. We can conclude from this calculation that, diffusion would be a significant process for moving substances within cells, but not for movement from shoots to roots. Experiments show us that the movement of photosynthates from shoots to roots can occur at rates of up to 100 cm h^{-1}.

use of aphids to study transport in phloem

Most plant scientists today consider that compounds are transported in the phloem by the mass flow of the solution. This was first proposed because of the fact that when phloem is cut an exudation forms at the cut surface. This was supported by work from a surprising quarter, involving aphids. Aphids, commonly known as greenfly or blackfly, feed on plant stems and leaves by inserting their feeding tube (stylet) directly into a sieve cell.

If a feeding aphid is anaesthetised and its stylet carefully severed leaving a short stump protruding from the stem, solution will exude from it at a rate of approximately 1μl h^{-1} for several days. Evidence that the stylet is tapping a phloem sieve cell comes from experiments using incisions into the plant tissue, designed to sever a sieve tube. Results of this type of experiment are shown in Figure 5.2.

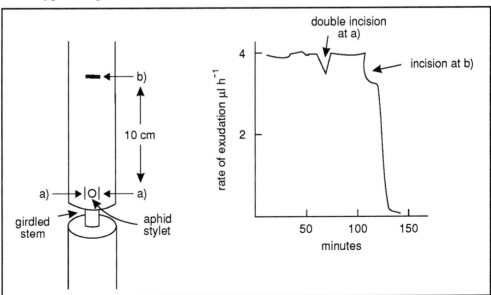

Figure 5.2 Effect of incisions into the phloem on the rate of excised aphid stylets on yound willow stems (after Weatherley, Peel and Hill, 1959, J Exp. Bot., *10*, 2-16). (See text for details).

Lateral incisions on either side of the aphid stylet (at a), in Figure 5.2) caused only a temporary perturbation of exudation, whereas an incision into the phloem directly above the stylet (at b) in Figure 5.2) stopped exudation completely. Interestingly, severance of the phloem more than 15 cm above the stylet was virtually without effect, which could be interpreted as suggesting that lateral movement occurs between sieve cells.

the mass flow hypothesis

The mechanism by which the mass flow may occur was first formulated by Münch in 1930. He proposed that sucrose, produced in the leaves as a product of photosynthesis, was transported into the sieve cells of the minor veins, where it accumulated to a high concentration. Water would move into the sieve cells by osmosis and a high turgor pressure would be generated as a result. In the root, sucrose is transported out of the sieve cells into the surrounding tissue where it is utilised in general metabolism and in growth. This results in the lowering of the turgor pressure in the root sieve elements. Münch proposed that, as a result of these differences in pressure, fluid would move in the sieve tubes by mass flow from the region of high hydrostatic pressure to the region of low hydrostatic pressure. Solutes would simply be carried along with the stream. The Münch pressure flow theory is outlined in Figure 5.3.

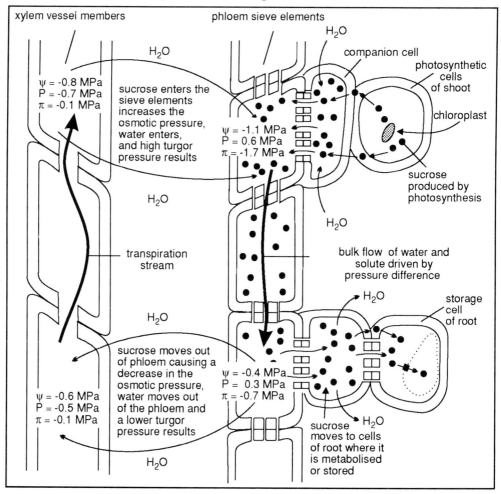

Figure 5.3 Diagram to illustrate the pressure flow theory of phloem operation. Possible values for ψ, π and P are indicated. Adapted and redrawn from Taizih and Zeiger, E (1991) Plant Physiology, Benjamin Cummins, New York.

The figures for ψ, π and P shown in Figure 5.3 have been obtained in a variety of ways. Early methods measured ψ and π, and calculated P by difference. More recent measurements determined P directly using either a micromanometer or pressure transducer sealed over the end of an aphid stylet. This is, not surprisingly, a tricky technique and its success is a tribute to the skill of the research scientists using it. The benefit of this method is that it provides a direct measurement of the pressure within an individual sieve element. Thus, there is now very good evidence for the occurrence of pressure differences between phloem in leaves and in stems a considerable distance away. The demonstration of a pressure gradient, of course, does not prove that the theory operates. It has, however, been tested directly using bean seedlings (Figure 5.4).

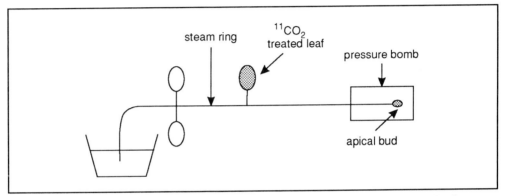

Figure 5.4 An experimental design to test the involvement of pressure differences in translocation (see text for explanation).

In this experiment, a leaf was exposed to $^{11}CO_2$ and produced radiolabelled sucrose. The steam ring prevented photosynthate moving down into the roots, so increasing the amount transported to the apical bud. The pressure bomb allowed pressure to be exerted around the apical bud thus modifying any gradient of pressure between the leaf and the bud. $^{11}CO_2$ was used in this experiment because even though it has a much shorter half-life than ^{14}C, it releases radiation which will pass through the walls of the pressure bomb and be measured by a suitably placed detector. Disintegration of ^{14}C releases β radiation which cannot penetrate the pressure bomb. Thus, $^{14}CO_2$ cannot be used in this type of kinetic experiment. Typical results are shown in Figure 5.5.

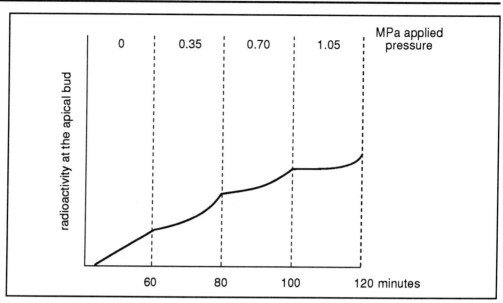

Figure 5.5 Diagram showing the effects of pressure around the apical bud on the transport of radiolabelled assimilate to the bud using an experiment based on the design shown in Figure 5.4.

Figure 5.5 shows that the imposition of 0.35 and 0.70 MPa pressure temporarily stopped the import of label into the bud, and that 0.70 MPa caused a longer stoppage than 0.35 MPa. Note that it took almost the full 20 minutes for recovery to begin after imposition of 1.05 MPa pressure. These results are interpreted as being in favour of the pressure flow theory, but try the following SAQ before we decide finally.

SAQ 5.2

Before accepting the data in Figure 5.5, there are one or two checks that ought to be done to ensure that we were not seeing artificial results.

What checks should be done?

Such checks have, of course, been carried out. Returning the pressure to zero after the 1.05 MPa treatment resulted in a full recovery of the import of photosynthate into the apical bud. Indeed, it is possible to cause the rate of import to fall and rise repeatedly by repeated application and release of 0.70 MPa pressure. Experiments to test the system to destruction, as it were, showed that 1.4 MPa pressure was required. Thus, the effects of up to 1.05 MPa pressure are considered to be valid and to support the pressure flow theory.

SAQ 5.3

An experiment was conducted to determine the effect of light, darkness and sucrose application on the translocation of the synthetic herbicide 2,4-dichlorophenoxyacetic acid (2,4-D). 2,4-D was applied to the first trifoliate leaf of a bean plant. If 2,4-D is translocated from the leaf into the stem, the stem shows a growth curvature due to the asymmetric distribution of the herbicide. Thus stem curvature shows whether or not translocation of 2,4-D has occurred. The results of this experiment are shown in Table 5.2.

How may these results be interpreted?

Treatment	Stem curvature (°)
light	0.0
light + 70 µg 2,4-D	31.0
dark	0.0
dark + 70µg 2,4-D	0.0
dark + 10% sucrose	0.0
dark + 10% sucrose + 70 µg 2,4-D	20.4

Table 5.2 The effects of veinous treatments on stem curvature (see SAQ 5.3).

5.4 Translocation occurs from sources to sinks

We have seen that the leaf is a source of sucrose and that pressure will build up in the phloem of such a region. Conversely, the root removes and utilises sucrose and is referred to as a sink.Thus, the pressure flow theory proposes that translocation occurs from a source to a sink.

SAQ 5.4

What other sinks operate in an actively growing plant? Make a list of as many as you can.

source-sink relationships

You should have been able to think of several alternative sinks but you may well find it difficult to think of alternative sources. Obviously, the leaf is the major source, but all of the storage organs will act as sources when development recommences. Thus, the potato tuber will be a source when it begins to sprout and the other storage organs will too. Even the young leaves of an apical bud are only temporary sinks. Once they reach a certain size they too begin to export sucrose. As you can imagine, this transition from sink to source is a fascinating one and is attracting considerable attention from plant scientists because of the information which might be revealed about how translocation is controlled.

5.5 Objections to the mass flow concept

Opponents of the mass flow theory have raised a number of objections to its applicability to all tissues. Münch's original theory maintained that the sieve elements were passive conduits through which solution flowed. There is considerable evidence that this is not the case. Thus, Bieleski showed that isolated phloem tissues of apple were able to accumulate sucrose against a concentration gradient. Other scientists have shown that inhibitors of respiration, such as KCN, inhibit transport in the phloem. Thus, modern scientists believe the phloem to be a highly metabolically-active tissue, but which still functions by mass flow.

A more difficult objection to answer is the fact that there are several pieces of evidence which show that solutes can move simultaneously both upwards and downwards in the phloem. An example of this is shown in Figure 5.6.

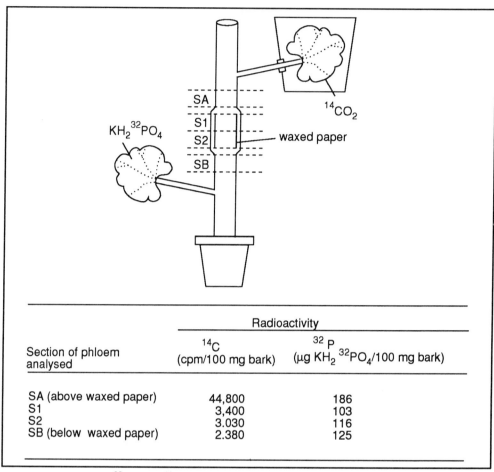

Section of phloem analysed	Radioactivity	
	^{14}C (cpm/100 mg bark)	^{32}P (μg KH$_2$ ^{32}PO$_4$/100 mg bark)
SA (above waxed paper)	44,800	186
S1	3,400	103
S2	3.030	116
SB (below waxed paper)	2.380	125

Figure 5.6 Movement of $^{32}PO_4$ and ^{14}C-labelled assimilates in the phloem of *Geranium*. Note that ^{14}C containing products had moved down into the experimental zone, whereas $^{32}PO_4$ have moved up, the implication being that the radiolabelled components are moving in the opposite direction to each other in the experimental zone. Data from Chen (1951) American J. of Botany 38,203.

directional phloem transport

In all of the experiments showing bidirectional flow, the phloem was treated as a whole tissue and was not subdivided in any way. Thus, there has been no demonstration of the bidirectional flow in a single sieve element. These experiments are interpreted by plant physiologists as showing that the entire phloem does not act as a single entity but that it is functionally subdivided, resulting in what appears to be the independent operation of the various parts. Unfortunately, plant scientists have not so far been able to discern the structural organisation of these independent units. One of the reasons for this is that the phloem is a very sensitive tissue and needs to be disturbed only a little for it to lay down callose in its sieve plates and to stop transporting. This is rather like being sent to study the bus system of a particular city but finding that the buses stop whenever you study them. Aphids are able to tap into the phloem without the system ceasing to function but so far we have not been able to find out how they do it. If we knew how aphids achieved this, perhaps we could use a new approach to phloem function. For now, the evidence for bidirectional flow described above is accounted for in terms of independent phloem sub-units, which consist, presumably, of several sieve tubes working in concert. However, there is evidence which seems to suggest that bidirectional flow can occur in one and the same sieve tube. The experimental design is shown in Figure 5.7.

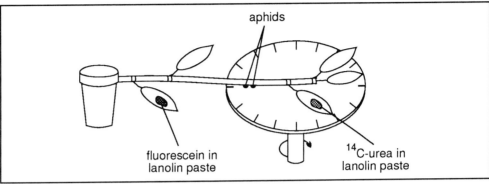

Figure 5.7 Demonstration of bidirectional translocation in the phloem of *Vicia faba*. Aphid exudate was collected on a plate revolving at a rate of 1 revolution in 24h. (See text for a description).

The aphids used in this experiment were of the species *Acyrthosiphon passim* which are known as a group to pierce and feed on the contents of a single sieve tube. Despite this the exudate from both aphids contained both ^{14}C-urea and fluorescein. Initially, this evidence seems to point to bidirectional transport in a single sieve element. However, it is known that lateral connections occur between sieve tubes and Figure 5.8 shows an explanation of the data.

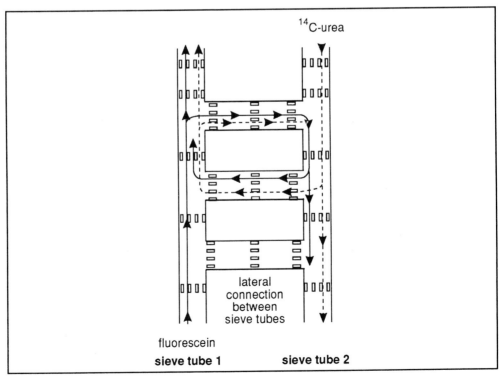

Figure 5.8 Lateral transfer and loop path as an alternative explanation for apparent simultaneous bidirectional movement in a single sieve tube (redrawn from Eschrich, 1967, Planta, **73**, 37-49).

The consensus among plant physiologists is that simultaneous bidirectional flow in a single sieve tube does not occur or, at least, it has never been conclusively demonstrated.

5.6 The problem of P-protein

The lumen of the sieve cell contains large amounts of a special protein called P-protein. P-protein is fibrous in nature, each fibril being approximately 5 μm long as far as can be ascertained from electron micrographs. Figure 5.9 shows a stylised representation of the deposition of P-protein in sieve tubes.

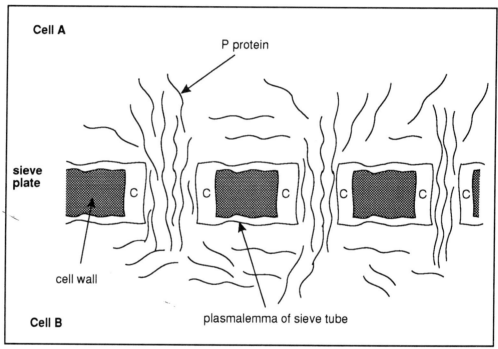

Figure 5.9 Stylised drawing of thin section through a sieve plate showing P-protein fibrils and callose (C) in the sieve pores. The magnification is in the region of x 10,000.

P-protein has featured from time to time in hypotheses suggesting the operation of muscle-like contractile systems which would somehow pump the fluid through the sieve tube. These hypotheses have foundered through lack of evidence but the P-protein remains as an enigma. Why should the sieve cell produce so much protein if it does not have a function?

5.7 How does sucrose get into and out of the phloem?

Having discussed the way in which sucrose is transported in the phloem we will now examine the processes referred to as phloem loading and unloading.

phloem loading   Sugars are manufactured in mesophyll cells of the leaf and are transported to the veins for movement to sinks. Minor veins play the major part here. In sugar beet, for example, minor veins are 13 times as extensive as major veins and each 10μm length of minor vein services eight or nine mesophyll cells.

The movement of sucrose in leaves can be determined using a combination of ^{14}C-sucrose and autoradiography. By supplying leaves for a period (say 30 minutes) with ^{14}C-sucrose, followed by washing and exposure to unlabelled sucrose for varying periods, it can be shown that sucrose leaves the mesophyll cells and enters the minor veins, subsequently, the sucrose enters the major veins.

In sugar beet and some other species such as pea, movement from the mesophyll cells to the minor vein is thought to involve an apoplastic step prior to loading into the sieve cell-companion cell complex. Before we proceed further try the following intext activity.

Π Remind yourself, if necessary, of the nature of the apoplast and symplast (Chapter 1) and then make a list of the sort of evidence you might look for when trying to decide whether or not there was an apoplastic step prior to sieve tube loading.

The evidence is as follows:

• there is a very low number of plasmodesmata linking the mesophyll cells with the sieve tube-companion cell complex (STC), but large numbers between the sieve tube and the companion cell themselves;

• there is a much higher solute concentration in the STC than in mesophyll cells and phloem parenchyma cells, a condition difficult to maintain in the presence of large numbers of open plasmodesmata;

• there is selective transfer of metabolites into the STC;

• apoplast exudates collected from leaves fed with $^{14}CO_2$ contain ^{14}C-sucrose.

We have given a very full list for completeness but we would have expected you to get the first three and maybe the fourth.

transfer cells   In the species showing the above characteristics there often occurs an unusual type of companion cell which is the initial site of uptake into the STC system. The unusual cell is called a transfer cell and it is specialised by having wall ingrowth protruding into the cytoplasm, which very significantly increases the surface area of the plasmalemma (Figure 5.10).

Transfer cells are agents for short distance transport and are found in strategic locations in the plant where such transport is important. These include the aleurone layer of cereal seeds, which transfers nutrients from the parent to the developing endosperm, and also the cells of salt glands which secrete salt.

In the transfer cells the increase in area of the plasmalemma increases the sites for insertion of the sucrose absorbing system, which consists of a proton pumping ATPase and a sucrose-proton symport (Figure 5.11). Effectively, this system uses the energy released by ATP hydrolysis to pump H^+ ions across the membrane. The build up of H^+ ions in the apoplasm is then used to drive sucrose uptake into the cytoplasm.

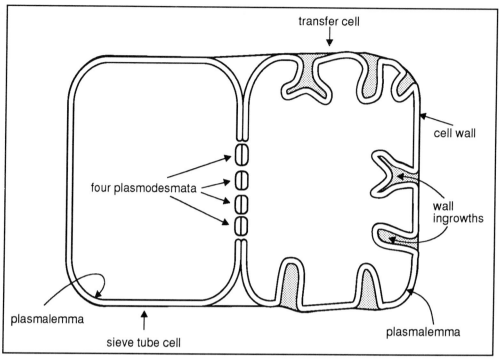

Figure 5.10 Diagram of transfer cell. Note the wall ingrowths, the convoluted plasmalemma and the lack of plasmodesmata except on the sieve tube face. Transfer cells also possess dense cytoplasm with abundant mitochondria.

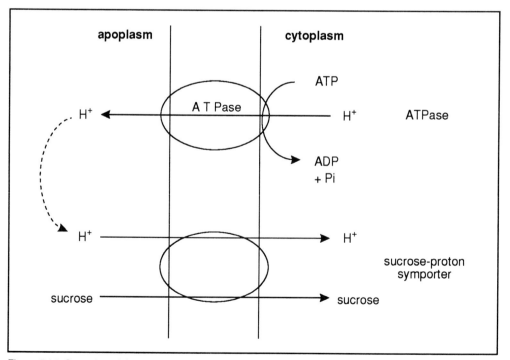

Figure 5.11 Operation of the sucrose-proton symport.

symplastic
route used in
some plants

Whereas there is very good evidence for apoplastic loading in sugar beet and peas, this is not universally the case. In numerous other species, including cucumber and maize, the evidence for apoplastic loading is either unclear or lacking. In these systems, there are numerous plasmodesmata linking mesophyll and STCs and only small concentration differences between them. There is little discrimination between metabolites transferred and apoplastic exudates from $^{14}CO_2$-fed leaves do not contain radiolabelled substances. We are forced to the conclusion that sucrose is directed to the STC in maize and cucumber by the symplastic route. This being the case, it is difficult to see how the STC could develop a higher turgor pressure than the mesophyll cells. High pressure in these systems presumably occurs throughout the whole of the source leaf and not simply in the STC itself.

5.7 Phloem unloading can also occur by two routes

phloem
unloading

There is good evidence that sucrose delivered to a young growing leaf is transported via the symplasm. This is also the case for root and shoot apical meristems. In these tissues there are numerous plasmodesmata linking the STC with the receiving cells.

SAQ 5.5

The compound p-chloromercuribenzene sulphonic acid (PCMBS) binds to the sucrose-proton symporter and prevents its activity. Experiments were carried out to determine the effect of PCMBS and, separately, of plasmolysis, on sucrose transport to root tips. What do you predict would be the effect of these treatments if sucrose was delivered to the root apex through the symplasm?

In tissues showing symplastic unloading, the sucrose is quickly utilised in the sink partly to provide respiratory substrates and partly to provide substrates for growth processes. This reduces the sucrose concentration in the phloem in the sink region thereby fulfilling the requirements of the pressure flow theory.

In tissues showing symplastic unloading once the sucrose has left the phloem it has entered the sink cells so there is no such thing as sink loading. This is not the case when apoplastic unloading occurs. The most closely studied examples of apoplastic unloading occur in developing seeds. Here the developing embryo belongs to a separate generation from the cells of the parent and there are no direct cytoplasmic connections between the two. Sucrose is delivered to the seed via the phloem and enters the seed by the vascular system which runs through the seed coat. Sucrose moves symplastically from the sieve cells into the cells surrounding the phloem until it reaches the inner margin of the seed coat (Figure 5.12).

Sucrose is released from the inner margin of the seed coat into the apoplast which, in cereals, is characterised by the presence of a cavity within the endosperm, the endosperm cavity. In legume seeds there is no endosperm cavity *per se*, simply a gap between the maternal tissue and that of the young seed.

Movement of sucrose into the developing seed occurs by absorption from the apoplast into the symplast of the seed. Thus, there is a definite sink-loading process here, in addition to the phloem unloading process.

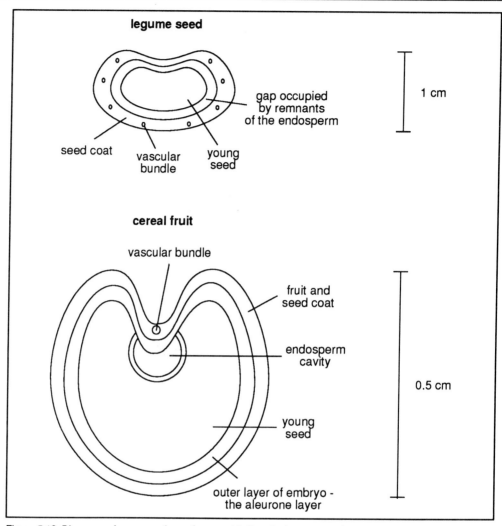

Figure 5.12 Diagrams of cross-sections of a cereal fruit and a legume seed.

SAQ 5.6

Describe an experiment which would enable you to decide if sucrose loading in cereal and legume seeds involved the operation of a proton-sucrose symport.

Once inside the symplasm of legume or cereal seeds, sucrose does not increase in concentration but is quickly converted into the fats, protein and starch stores of the seed. This is also the case with many vegetative storage organs such as the potato tuber. Sugar cane and sugar beet are interesting because, as you know, they store sucrose and between them satisfy the world's demand for this commodity. In these two species there are no plasmodesmata linking the STC with the storage cells. Sucrose is considered to be transported out of the STC symplasm and into the storage cell symplasm through facilitated diffusion but is transferred into the vacuole by the operation of a proton-sucrose symport, located on the tonoplast. Sugar cane and sugar beet are very important as food sources and are receiving close attention in an attempt to unravel the details of the control of sucrose accumulation.

Having examined how sucrose is transported from source to sink and how phloem loading and unloading occurs let us now examine the situation in the plant as a whole.

5.8 Plants have a hierarchy of sinks

If experiments are conducted to determine the import or export of photosynthate in a plant, results similar to those shown in Figure 5.13 are commonly obtained.

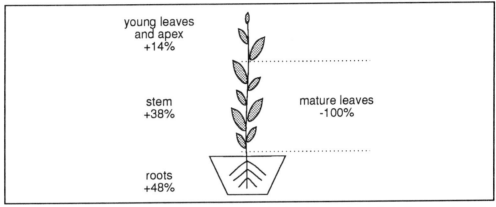

Figure 5.13 Import and export of photoassimilate in a whole plant. The numbers refer to the percentage of photosynthate transported, + implies import, - implies export.

This type of experiment has been done many times and in several ways. It was done first by determining the changes in mass of leaves, stems and roots in experiments using large numbers of carefully selected replicates. More recently it has been done by assessing the distribution of radiolabelled assimilate. For plants growing vegetatively the results are always the same. The mature leaves are net exporters and are the major sucrose sources in the plant. The roots and the young parts of the shoot are importers (sinks), as would be expected, but so too is the stem, even though it is green in most species. The stem and young leaves together receive a higher proportion of photosynthate than the roots. Within this distribution pattern the upper mature leaves transport material to the young leaves and apex, whereas the lower ones transport assimilate to the roots. Intermediate leaves distribute products in both directions. As mentioned above this is the situation in a plant growing vegetatively. When reproduction begins photosynthate distribution changes markedly.

changes in the distribution of sinks during reproduction

The formation of the flower, and later the seed and fruit, creates several new sinks. The flower itself is no more attractive of photosynthate than the young leaves and apex, but the seeds and fruit are, so much so that roots receive very little material and virtually stop growing. In plants which produce underground storage organs, such as potato, the seed and fruit are insignificant and the underground organ develops the same attractive force as seeds and fruits in other species. We can, therefore, recognise a hierarchy of sinks which changes as the plant progresses through its developmental cycle. This phenomenon is extremely important to mankind because the delivery of sucrose to sinks controls their increase in mass and governs the yield of the harvested part of the plant. If we can control this process we might be able to control the growth of the storage organs and, thereby, help to solve the world's food problems. What controls the attractive ability of sinks? We will attempt to answer this question in the next section.

5.9 The ability to attract assimilates is termed sink strength

Needless to say we do not yet know the complete answer to the question posed. Several pieces of evidence suggest that the ability of a sink to attract assimilates, referred to as sink strength, depends upon two factors, sink size and sink activity, such that:

$$\text{sink strength} = \text{sink size} \times \text{sink activity} \qquad \text{(E-5.1)}$$

sink strength

Sink size is relatively easy to understand but there is disagreement as to what parameter of size is important. Thus, size is recorded by some researchers as its mass, while others consider that cell number is a better measure. The value of the latter is demonstrated by evidence for a positive correlation between final fruit mass and cell seed number. This relationship was shown first in cereals but is also known to be the case in many legumes and in sunflower. Sink activity can be considered to refer to the sum total of the metabolic activities of the sink and is a reflection of the sink's ability to absorb and utilise sucrose.

The fact that overall sink size is important can be demonstrated very easily. For example, in experiments in which grain was removed from the developing wheat ear, lower final mass for the ear was produced. Closer examination of the experimental ear revealed an intriguing situation. Each grain in the experimental ear was heavier than in the control. Thus, although total mass was reduced, there had been some compensatory activity. This is common in many crops, especially oilseed rape, but cannot easily be explained by the equation relating sink strength to sink size and activity.

changes to source activity

Further, evidence suggests that source activity can also change and this may have an effect. When all but one of the source leaves on a soybean plant are shaded for an extended period, such as eight days, the sucrose synthesising and exporting ability of the exposed leaf increases dramatically. Thus, source activity itself is not invariate.

There is one other line of evidence which suggests that Equation 5.1 is too simple: this involves experiments in which the sink end of the phloem pathway is bathed in a solution containing a high concentration of salts.

These experiments were carried out first in an ingenious system, using what are referred to as half-seed coat cups (Figure 5.14).

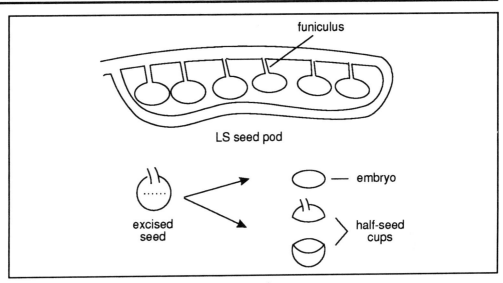

Figure 5.14 Production of half-seed cups from bean seeds.

<table>
</table>

experiments
with half-seed
cups

Bean fruits can be opened to reveal the developing seeds each attached to the fruit wall by a funiculus. The seed can be excised and the seed coat carefully cut, thus allowing the removal of the embryo and providing two half seed coat cups. If the parent plant is fed $^{14}CO_2$, ^{14}C-sucrose is generated and transported into the seed where, as we saw above, unloading is apoplastic. Excision of half-seedcoat cups from such plants provides material which contains considerable quantities of ^{14}C sucrose and tests show that the cup will continue to unload ^{14}C sucrose for some considerable time. Solutions can be placed inside the cup and experiments conducted to determine the effect of solution constitution on unloading. These experiments demonstrate that unloading is dramatically stimulated by solutions of high solute concentration, such as 1.0 MPa, compared to water or solutions of lower concentration. Tests on intact bean seeds and cereal grain show that concentrations of solute such as this are normal in the apoplast of developing seeds. This effect of solute concentration on phloem unloading has also been demonstrated in grape berries and in a totally different system, the root.

These experiments suggest that sink activity should include the ability not only to absorb and metabolise sucrose, but also to release metabolites so as to increase the solute concentration of the sink apoplast. This evidence may also suggest the need to re-examine the pressure flow theory as originally stated. The theory explained the production of pressure differences at the two ends of the phloem conduits entirely by differences in sucrose concentration. This need not be the case. As long as a pressure difference is generated there will be a mass flow of solution in the phloem, and sucrose will be carried along with it. The experiments reported above can be interpreted as causing a reduced turgor pressure in the sink phloem which will increase the pressure gradient between source and sink and result in an enhanced delivery of sucrose. This effect of solutes in the sink must be considered to be an additional component and, because of its importance in contributing to the control of yield, is the topic of very active research.

5.10 Sucrose is not the only compound transported in the phloem

The composition of phloem sap is shown in Table 5.3.

Component	Concentration (g l^{-1})
sugars	80-100
amino acids	5
protein	1.5-2.0
chloride	0.5
phosphate	0.5
potassium	3.0
magnesium	0.15

Table 5.3 Composition of phloem sap in castor bean.

other phloem constituents

Whereas sugars constitute the major part of the solutes transported in the phloem there are significant amounts of other substances, particularly amino acids and potassium ions.

The mineral ion content of the phloem varies both with the zone of the plant and the time of year. While minerals are transported into leaves via the xylem there is a considerable re-mobilisation of minerals from older to younger leaves. This movement occurs in the phloem. There is an even greater mobilisation in perennials associated with the preparation for leaf fall in Autumn.

SAQ 5.7

The phosphate pool in a leaf can be made radioactive by incubating the leaf with $^{32}PO_4$. Design an experiment to find out if the re-mobilised phosphate from an ageing leaf is transported in the xylem or phloem.

Experiments of the type described in SAQ 5.7 show that the mobilisation of minerals from ageing leaves occurs in the phloem.

We have now studied what may be considered to be the major physiological processes of plant life. Our theme has been to point out the consequences of terrestrial life and to describe the ways in which plants overcome these and survive. Survival, of course, also involves growth and reproduction and the process of flowering and fruiting. Before we examine these phenomena, however, we need to know something about plant hormones, and that is the topic of the next chapter.

Summary and objectives

Numerous pieces of evidence show that sucrose is transported around the plant in the phloem and the most widely accepted theory to account for this is the Münch pressure flow theory. There is direct and indirect evidence in favour of the theory and studies using sap-sucking aphids have contributed to this. Movement is from sources to sinks and plants contain numerous sinks which compete for materials being translocated. Compounds enter the transport stream by phloem loading and leave it by phloem unloading. In some instances these processes involve apoplastic steps and in other symplastic steps. Bidirectional flow can occur in the phloem as a whole, but not in a single sieve element. Compounds other than sucrose are transported in the phloem and it is the site of transport of re-mobilised minerals.

Now that you have completed this chapter you should be able to:

- show an understanding of the pressure flow theory and interpret supplied data designed to test the theory;

- show an understanding of the terms phloem loading and unloading and list the evidence which would enable a distinction to be made between apoplastic and symplastic routes for these processes;

- describe the meaning of the term sink strength and explain its importance to phloem function and plant productivity;

- design experiments to determine the route of movement of sucrose and of remobilised minerals.

Plant hormones

Plant hormones

6.1 Introduction

It is always difficult to decide when to discuss plant hormones in a plant physiology course. We have decided to insert them as Chapter 6 of our text, after the largely physiological material but before a consideration of growth and developmental phenomena. The aim of this chapter is two-fold. One is to describe the physiology and biochemistry of hormones as a prelude to chapters on reproduction (Chapter 7), and especially on growth and development (Chapter 8). Secondly, whilst recognising that biotechnology seeks to manipulate plants, and that this text underpins many of the other texts in the BIOTOL series, it is important to know something about how hormone levels might be artificially manipulated.

∏   In this chapter, we will use many abbreviations as a simplified way of writing about some quite complex chemicals. These abbreviations are commonly used by plant physiologists so it is important that you know them. We would suggest that you make a list of these as you go, noting the activities of each. We suggest you use the following format:

Abbreviation	Chemical name	Activities

6.2 What is a hormone?

The term hormone was coined in the 1920s to describe the action of low molecular weight compounds produced by animals, which have the following characteristics:

- they are produced at a specific site;

- they are transported from a separate site of production to a site of action;

- they bring about specific identifiable changes at the site of action.

A number of compounds are known to be produced in plants which affect growth, differentiation and certain physiological processes, and these compounds have some but not all of the above characteristics of hormones. For this reason the compounds are often referred to as plant growth regulators (PGRs). The main obstacle to these compounds qualifying as hormones is the fact that we do not yet know what specific changes they bring about at the site of action. There is also a problem with identifying a sufficiently localised site of synthesis, since plants have incredible powers of regeneration and so can regenerate leaves, branches and roots from almost all of their parts. This has, of course, been instrumental in plant survival. Indeed, browsing and grazing animals would soon consume all plant material if the plants were unable to regenerate.

There is clear evidence for the involvement of PGRs in the regeneration of plant organs which appears to show that the PGRs are made within the regenerating parts. This does not comply with the first two requirements of a hormone and so some scientists have concluded that plants do not produce hormones. However, simply because wounded and regenerating parts produce PGRs, does not mean that all parts of healthy, intact plants produce them also, and hence it is possible that the first two characteristics of hormones are met in intact plants.

There is one other aspect of the naming of compounds that we should address. We will see below that a number of synthetic compounds mimic the effect of naturally-occurring PGRs and to call both groups by the same name may be confusing. This problem has been avoided by referring to the naturally-occurring compounds as hormones and to the synthetic compounds as PGRs. We will do the same.

6.3 Numerous factors affect hormone steady-state levels

A very important factor concerning a hormone is that its concentration must be controlled. This is to ensure that the rate of the processes which it affects are also controlled. The factors involved in this control are shown in Figure 6.1.

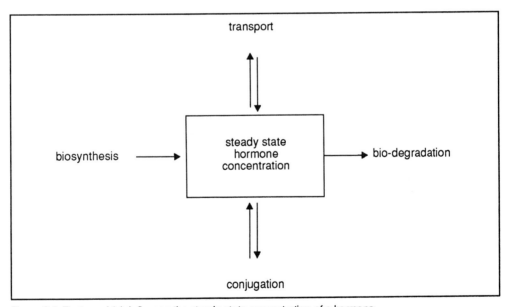

Figure 6.1 Factors which influence the steady-state concentration of a hormone.

hormone biosynthesis and degradation

conjugation

The most obvious factors are biosynthesis and biodegradation and these are both irreversible processes. Biosynthesis can only lead to an increase in concentration, whereas biodegradation can only lead to a reduction. Most plant hormones have been shown to form conjugates with sugars and/or amino acids. The formation of the conjugate appears to be a detoxifying system in some species because the formation is irreversible. In others, they act as temporary storage forms because their formation is reversible.

compartment-
ation

A less obvious factor is that of transport, which has two elements. At the intra-cellular level, a hormone may be moved from one compartment to another, such as from the cytosol to the vacuole. This phenomenon is called compartmentation and is often experimentally difficult to assess. Hormones may also move from cell to cell, ie they can be transported into or out of a particular cell. This is difficult to assess at the purely cellular level but useful information can be obtained concerning the differences in hormone concentrations between tissues. We will look at the ways in which hormones are measured as a prelude to the examination of the factors affecting their steady state concentrations.

6.4 Three techniques are used to quantify hormones

bioassays

Plant hormones can be assayed by:

- bioassay;

- immunologically-based methods;

- physico-chemical methods.

Bioassays. As we will see in later sections, plant hormones were discovered by the study of growth and differentiation phenomena and these initially provided the only way of assaying the hormone. Such a method is called a bioassay. In bioassays the amount of a compound is determined by measuring its effect on a biological response, such as the growth of part of an organism. Bioassays have both advantages and disadvantages. They are useful because they demonstrate the biological activity of a compound but they suffer because, in general, they are not absolutely specific for the compound in question and cannot be used for precise quantification.

We will illustrate the application of bioassays by describing the assay of indole-3-acetic acid (IAA) using oat coleoptiles.

Examine Figure 6.2 first before continuing with the text. The stem apex and the developing leaves are surrounded by the coleoptile, which is a more-or-less cylindrical organ . The seedling illustrated in Figure 6.2 has been grown in the dark for 4 days. In a seedling of this age, the stem is very short and its apex is situated just above the first node.

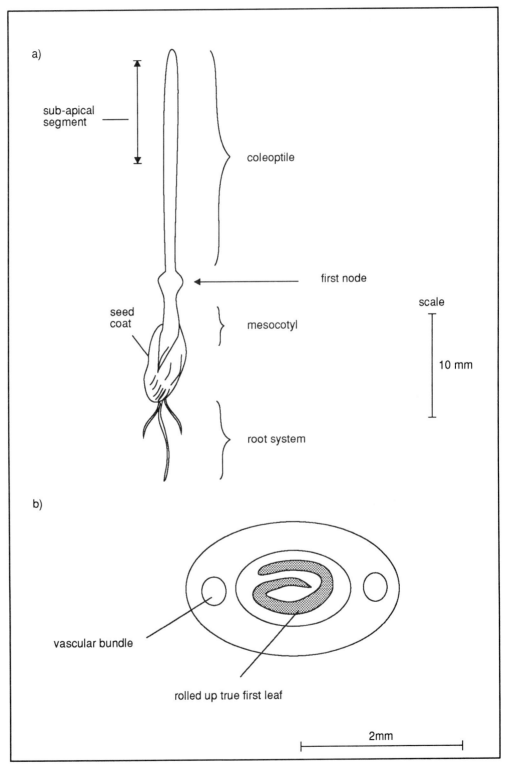

a)

sub-apical segment

coleoptile

first node

seed coat

mesocotyl

scale

10 mm

root system

b)

vascular bundle

rolled up true first leaf

2mm

Figure 6.2 Diagram of four-day-old oat seedling showing a) general morphology and b) a transverse section through the coleoptile illustrating its cylindrical nature.

If 10 mm sub-apical segments of coleoptiles are cut and floated on solutions of the plant hormone indole-3-acetic acid (IAA), they will grow in length. Table 6.1 shows the lengths of such sections after a 24 hour incubation on IAA solutions. (The structure of IAA is shown in Figure 6.7).

Replicate number	IAA concentration (mol l^{-1})						
	0	10^{-8}	10^{-7}	10^{-6}	10^{-5}	10^{-4}	10^{-3}
1	17	17	17	19	21	19	15
2	15	18	21	21	25	18	14
3	14	17	18	20	22	17	13
4	15	15	20	18	23	19	13
5	16	18	16	22	22	17	16
6	16	16	15	23	24	21	13
7	17	16	20	21	20	19	14
8	18	19	22	24	24	17	13
9	17	17	18	21	24	20	16
10	17	18	23	19	22	18	15

Table 6.1 Lengths (mm) of coleoptile segments after 24 hour incubation on solutions of IAA. The initial length of the coleoptile segments was 10 mm.

Ⅱ Calculate the average amount of elongation growth in each treatment. Express the results as a percentage of the control and plot the results against IAA concentration. Now determine the amount of IAA in an extract which caused a growth increment of 150%. When you have done this, check with the results in Table 6.2 and Figure 6.3.

IAA (mol l^{-1})	Average length (mm)	Average growth (mm)	Elongation growth as % of control
0	16.2	6.2	100
10^{-8}	17.1	7.1	114.5
10^{-7}	18.9	8.9	143.5
10^{-6}	20.8	10.8	174.4
10^{-5}	22.7	12.7	204.8
10^{-4}	18.5	8.5	137.0
10^{-3}	14.2	4.2	67.0

Table 6.2 Average results from Table 6.1.

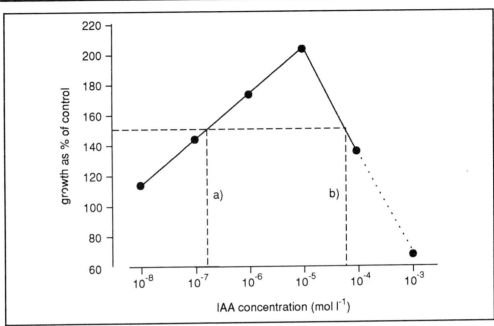

Figure 6.3 Effect of IAA concentration on elongation growth of coleoptile sections (see text for further explanation).

Note that between 10^{-8} and 10^{-5} mol l^{-1} there is a linear relation between elongation growth and the logarithm of the IAA concentration,whereas inhibitory results were obtained at higher concentrations of IAA. Two values are possible for a growth increment of 150%. Vertical lines from this position cross the x axis at position a) between 10^{-7} and 10^{-6}, and at position b) between 10^{-5} and 10^{-4}. Note the logarithmic scale. Hence the two values are approximately 2×10^{-7} mol^{-1} and 9×10^{-5} mol l^{-1}.

How would you decide which of the two results was correct?

optimal, sub-optimal and supra-optimal concentrations

This can not be done without further experimentation. The sample of unknown IAA concentration must be diluted by a known amount and the test repeated. If the lower figure is correct the diluted sample will stimulate growth to a lower amount (that is, less than 150% of control). With reference to Figure 6.3, the concentration producing the highest response is called the optimal concentration. Lower and higher concentrations are called sub-optimal and supra-optimal respectively.

immunological assays of plant hormone

Immunological methods rely on the ability of antibodies to recognise and bind to specific antigens. If antibodies can be generated against a plant hormone they will bind specifically to hormone molecules in plant extracts. The concentration of an antibody-hormone complex can be determined in various ways and this provides a measure of the concentration of the hormone in the plant extract. Plant hormones are too small to induce an antibody response by themselves. However, if they are covalently-linked to a macromolecule, such as bovine serum albumin (BSA), then the hormone-BSA conjugate acts as an antigen when injected into an animal such as a rabbit and induces the production of antibodies which react with the hormone. The binding of antibodies is so specific that they can distinguish between molecules which are structurally very similar. Antibodies may be used to assay hormones in plant extracts using radioisotope-labelled or enzyme-labelled antigens and these techniques have the advantage of being very sensitive and of reacting linearly rather than logarithmically with the concentration of the hormone. The technique is illustrated in Figure 6.4.

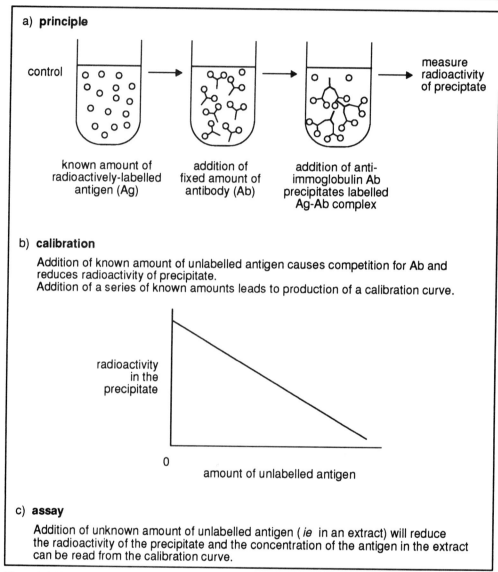

Figure 6.4 Use of radioimmunoassay to quantify hormones.

Use of HPLC, GLC and MS in the assay of phytohormones

Physico-chemical methods used in hormone analysis include various permutations of high performance liquid chromatography (HPLC), gas liquid chromatography (GLC) and mass spectometry (MS). HPLC and GLC are essentially mechanisms for separating compounds. In the area of plant hormone analysis, HPLC tends to be used to achieve partial purification of a sample. The putative hormone is then converted into a volatile derivative which is applied to a gas chromatograph linked to a mass-spectrometer (GCMS), a typical read-out from which is shown in Figure 6.5.

Such a spectrum as that shown in Figure 6.5 can be used to qualitatively identify a compound, but if you are confident that the compound is present, the MS can be set to detect only a single ion, in which case it can be used to quantify compounds with high sensitivity. If you are unfamiliar with mass spectrometry, we would recommend the BIOTOL text, 'Techniques Used in Analysis of Bioproducts'.

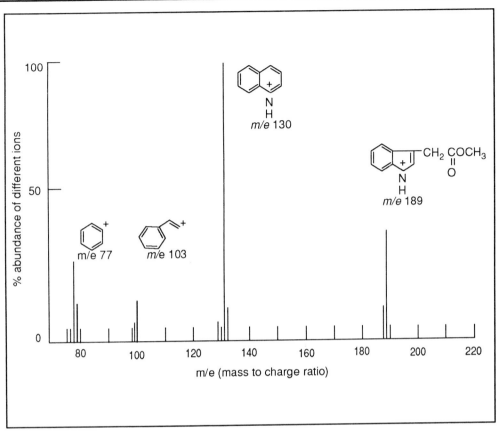

Figure 6.5 Mass spectrum of the methyl ester of IAA.

We are now ready to look at the different hormone classes.

6.5 Plant hormones are grouped into five main classes

It is usual to divide plant hormones into five classes, which partly reflects the order in which they were discovered but which also recognises differences in their chemical nature. The five groups are auxins, gibberellins, cytokinins, abscisins and the solitary gas, ethylene. The stimulation of fruit ripening by ethylene has been known for many years but it was not immediately designated as a hormone. We might ask how can the concentration of a gas be controlled? The detailed knowledge of its biosynthesis, which we will examine later in the chapter, reveals a stable precursor whose distribution and concentration can be controlled. Thus, ethylene is nowadays designated as a hormone. Towards the end of the chapter we will also mention other compounds which may or may not qualify as hormones, but which have important affects on aspects of growth and differentiation. We will now examine the five main groups in turn.

6.6 Auxins

Π   We will make similar lists of the physiological/morphological effects for the other groups of plant hormones. It would be sensible to make your own summary sheet of these so that you can make a direct comparison of the effects of the various groups of hormones. To do this, copy out Table 6.3 onto a clean sheet of paper and add the corresponding tables for the other hormones when you come to them in the text.

Physiological and morphological effects involving auxins

  i)   cell elongation in stems, coleoptiles and roots

  ii)   inhibition of growth of axillary buds in apical dominance (see Chapter 8)

  iii)   formation of lateral roots and adventitious roots

  iv)   delay of onset of leaf abscission

  v)   growth of fruit

  vi)   control of cambial activity

Table 6.3 The physiological and morphogenetic effects involving auxin.

Auxins were discovered as a result of observations made by Charles Darwin and his son Francis in the 1880s. They noted that if the coleoptile of canary grass (*Phalaris canariensis*) was exposed to light from one side the coleoptile bent towards the light. Experiments by these and other scientists showed that a growth-modifying compound was produced in the tip of the coleoptile that was transported to the subapical zone and there caused cell elongation to occur. This compound could be collected in an agar block which itself would cause bending when applied asymmetrically to a decapitated coleoptile. These observations are illustrated in Figure 6.6

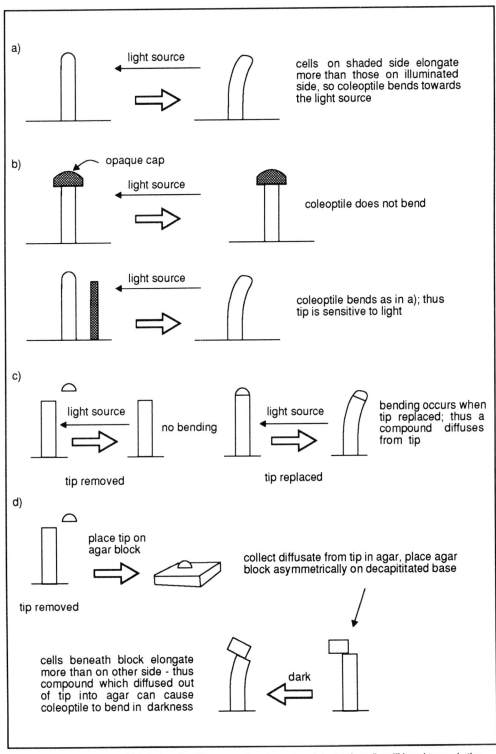

Figure 6.6 Experiments which led to the discovery of auxins; a) an intact coleoptile will bend towards the light; b) but only if the tip is exposed to light ; c) removal of the tip prevents curvature but bending is retained if the tip is replaced; d) an agar block containing a diffusate from an excised tip will cause bending in the dark if placed asymmetrically.

The compound causing these changes was isolated in 1934 and called auxin, from the Greek word meaning to grow, and a compound is said to have auxin activity if it is active in an auxin bioassay, such as the oat coleoptile test described earlier. Four naturally-occurring compounds are known which will do this (Figure 6.7).

Note that they are often referred to by abbreviations, which are useful to remember.

IAA, IBA and 4-Cl-IAA are equally active in bioassays and 10-100 times as active as PAA.

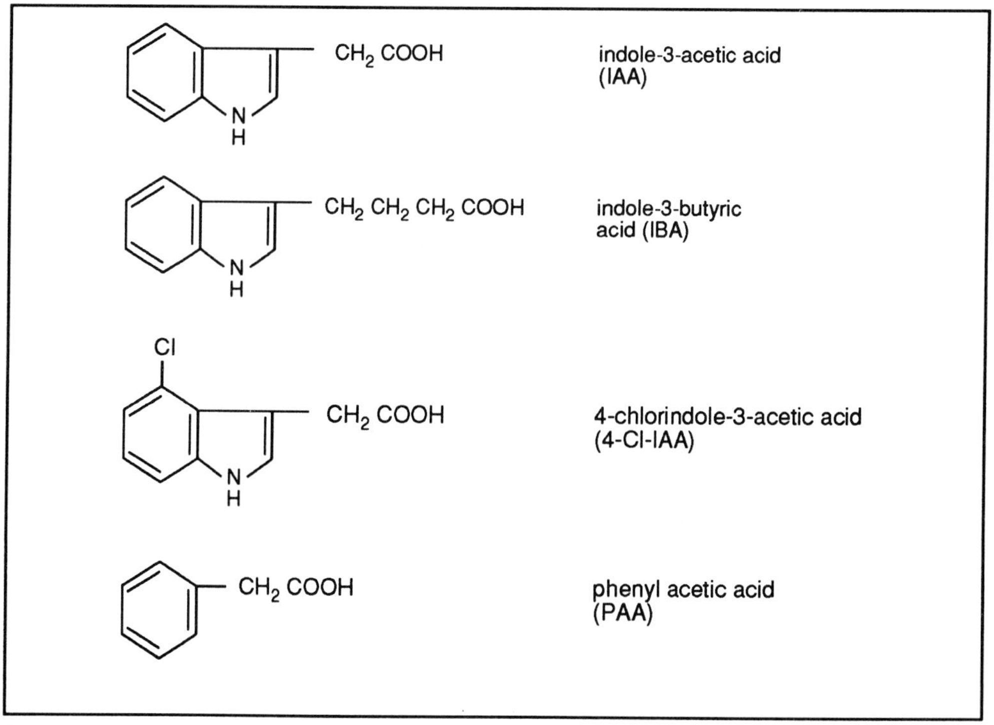

Figure 6.7 Structure of natural auxins.

6.6.1 Biosynthesis and degradation

IAA biosynthesis occurs in meristematic regions from the amino acid tryptophan and there are two pathways in plants for achieving this. (Figure 6.8); the IPA (indole-3-pyruvate) route appears to be the major one.

Indole-3-ethanol acts as a reservoir for excess IAald and can be reconverted to it. PAA is synthesised from the amino acid phenylalanine by a pathway analogous to the IPA route to IAA. The synthesis of IBA or 4-Cl-IAA has received little attention.

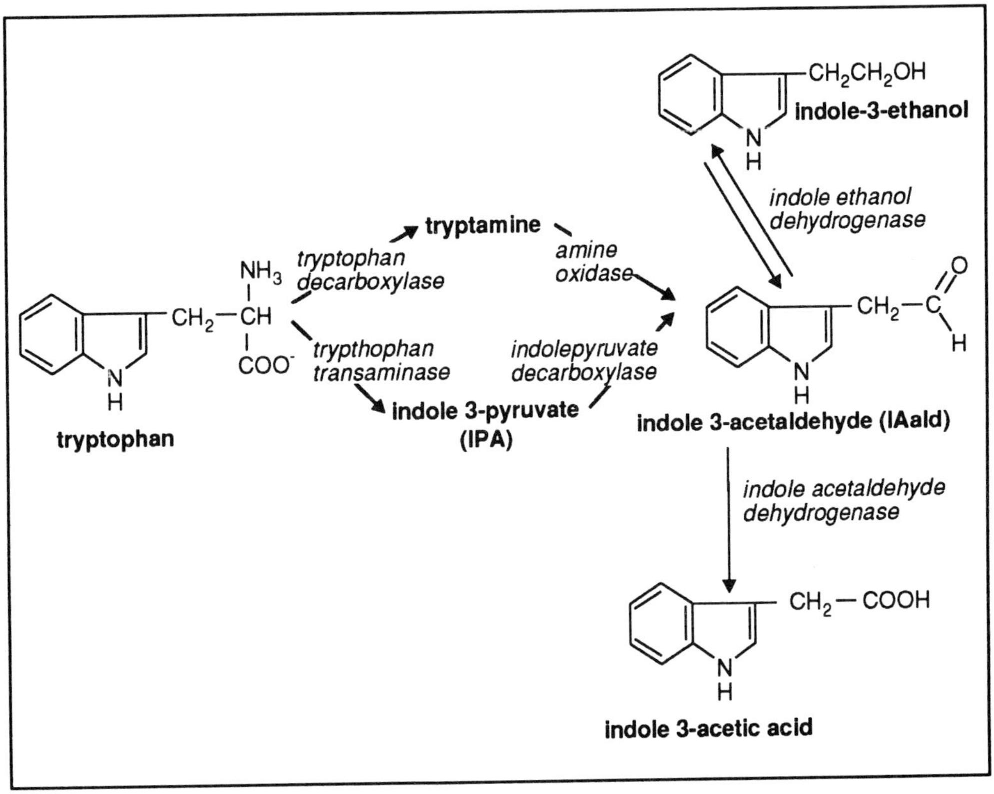

Figure 6.8 IAA biosynthesis from tryptophan.

IAA oxidase   IAA degradation is caused by a group of poorly characterised enzymes, generically
referred to as IAA oxidase, which includes a number of peroxidases. It is not surprising,
therefore, to find several IAA degradation products (Figure 6.9).

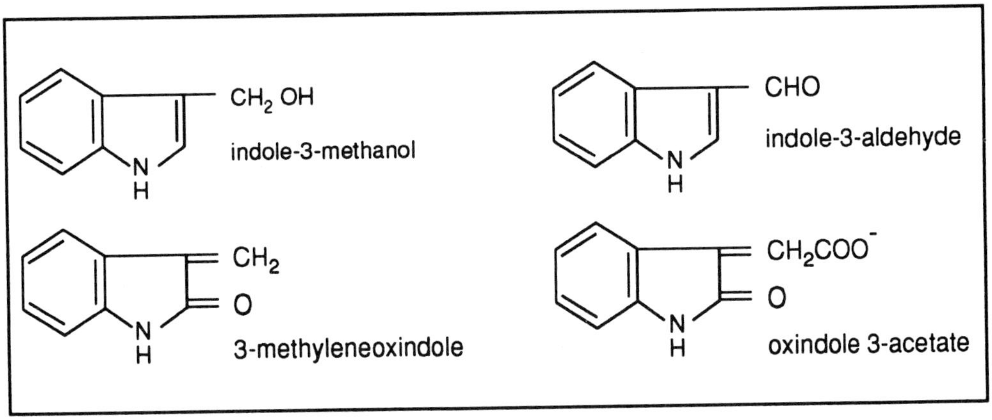

Figure 6.9 IAA degradation products.

SAQ 6.1   The compounds shown in Figure 6.9 are called degradation products which
implies that they do not show auxin activity. How would you test this?

In IAA degradation, once an oxygen atom has been introduced into the indole group the ring is much less stable and degradation subsequently produces small molecular weight compounds which are recycled through other metabolic pathways.

bound (conjugated) auxin

In addition to free IAA, plants also possess a number of covalently bound forms, called bound or conjugated auxins, which may be low or high molecular mass compounds. Examples of these are shown in Table 6.4.

Low molecular mass	High molecular mass
IAA - *myo*-inositol IAA - *myo*-inositol - arabinose IAA - *myo*-inositol - galactose IAA - glucose IAA - aspartic acid IAA - glutamic acid	IAA - glucan (7-59 glucose units) IAA - glycoprotein

Table 6.4 IAA conjugates.

role of conjugates

The low molecular mass compounds are either esters of sugars and sugar derivatives or amides of amino acids. Their role appears to vary from species to species. In some species, formation of conjugates is reversible and so conjugates may act as temporary stores of IAA. In others, their formation is irreversible and they are considered as part of the process of IAA inactivation. The IAA bound in high molecular mass conjugates is active only when released. IAA glycopeptides are found in cereal seeds where they are degraded during germination. Thus conjugates seem to act as stored forms of IAA which are apparently protected from degradation by IAA oxidase.

6.6.2 IAA is transported in two systems

It was indicated above that much of the early work on IAA was carried out using coleoptiles. This was also the case for the study of IAA transport and it utilised the donor-receiver agar block method (Figure 6.10).

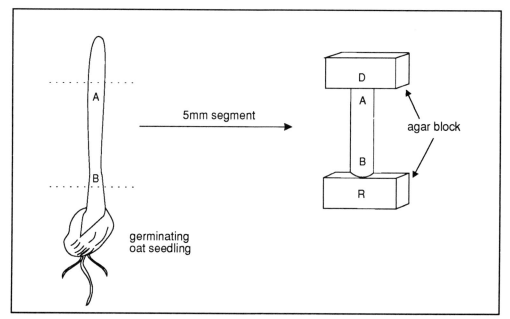

Figure 6.10 Donor-receiver agar block method for measuring IAA transport. A is the apical end and B the basal end of the coleoptile segment. IAA is contained within the donor block, D, and transported to the receiver block, R.

Numerous replicates of segments and blocks are arranged and a chosen number of receivers are removed at each time interval and their IAA content determined. The IAA was originally assayed by bioassay but in later studies radioactive IAA was used. When arranged as in Figure 6.10 the method measures the rate of movement towards the base of the coleoptile, ie in a basipetal direction. By inverting the coleoptile segment so that the basal end is adjacent to the donor block the rate of movement towards the apical end, ie in an acropetal direction, is measured.

SAQ 6.2

A series of coleoptile segments (average length 5.44mm) were set up to measure the basipetal movement of ^{14}C-IAA. Receiver blocks were removed at time intervals and their radioactivity assessed. The results are shown in Table 6.5.

Plot a graph of radioactivity against time and calculate the rate of IAA transport in mm h^{-1}. There is a hint in the response if you need it.

Time (hours)	Radioactivity (counts/min)
1	10
2	65
3	125
4	190
5	250

Table 6.5 Radioactivity in receiver blocks after ^{14}C-IAA transport (see SAQ 6.2).

Rates of basipetal transport of 5-15 mm h^{-1} are routinely found for coleoptiles, stems and petioles but acropetal transport is much slower, usually 0.5-1 mm h^{-1}. Thus IAA transport in the shoot is polarised in the basipetal direction. In roots the situation is the opposite, transport occurs at approximately 10 mm h^{-1} in the acropetal direction but at less than one tenth of this in the basipetal direction, ie polarised in the acropetal direction. This is not as strange as it seems since basipetal movement in the shoot is in longitudinal continuity with acropetal movement in the root. Thus, the polar transport of IAA will tend to transport IAA from the shoot into the root and there is good evidence that this actually occurs, especially in young plants. Figure 6.11 shows a proposed mechanism for this movement.

IAA is a weak acid and its dissociation is affected by pH. At apoplastic pH values it is undissociated (IAAH). Undissociated IAAH is lipophilic, and therefore able to cross lipid membranes. At cytoplasmic pH values, IAA dissociates to form the IAA anion (IAA$^-$), which is lipophobic and, therefore, cannot diffuse out. Under these conditions IAA, in the form of IAA$^-$, would accumulate. If efflux carriers specific for IAA$^-$ were present in the plasmalemma, IAA$^-$ efflux would occur down its concentration gradient. The location of such carriers towards one end of the cell would impose a polarity on the movement of IAA. There is evidence for such a distribution of IAA$^-$ carriers. This evidence comes from using compounds which inhibit IAA transport. Figure 6.12 shows two compounds which inhibit the polar transport of IAA.

polar transport of IAA

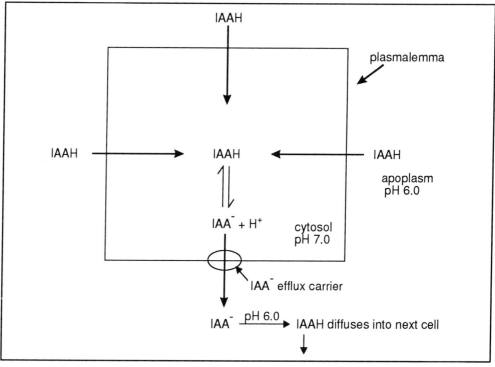

Figure 6.11 Chemosmotic model to describe the basis for the polar transport of IAA. (See text for a description of the process).

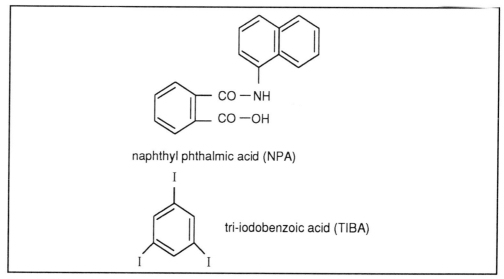

Figure 6.12 Structure of two auxin transport inhibitors.

These compounds both act by binding to the IAA efflux carriers, thereby preventing IAA-transport. Jacobs has managed to extract and purify this NPA-binding protein, raise antibodies to it and render these antibodies fluorescent by binding a fluorescent marker to them. Application of these antibodies to sections of dark-grown pea stem tissue reveals that the NPA-binding protein is concentrated at the basal end of the cells.

The exact cellular site of IAA moving in the polar transport system is not known but appears to be in phloem parenchyma. IAA applied to the leaves of mature plants is transported in the phloem sieve tubes but this distribution is not polar and is not inhibited by NPA or TIBA. IAA which is formed in senescing leaves, prior to leaf fall, is also transported in phloem sieve tubes. Thus IAA can be transported in two systems. We noted earlier that IAA biosynthesis occurs in meristematic areas, such as the apical meristems of shoots. Its primary distribution around the plant is considered to be by the polar transport system, with any subsequent redistribution taking place in the phloem.

SAQ 6.3	Figure 6.13 shows the results of the assay of the IAA content of the hypocotyl and epicotyl of a young dicotyledonous seedling. Interpret these results in terms of the likely site of IAA synthesis and degradation.

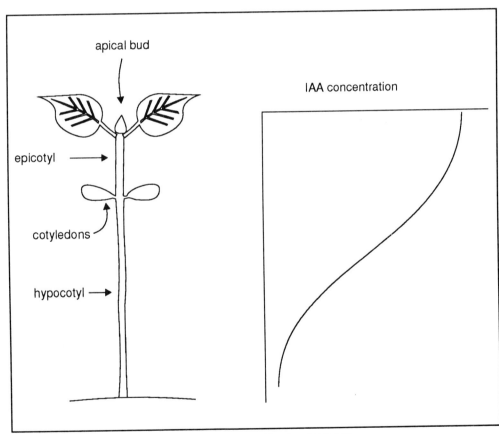

Figure 6.13   IAA distribution in the axis of a young seedling (see SAQ 6.3).

SAQ 6.4

Fill in the squares in the diagram below using: export; precursor; degradation; blocked by TIBA; temporary storage; biosynthesis. One of the phrases may be used twice.

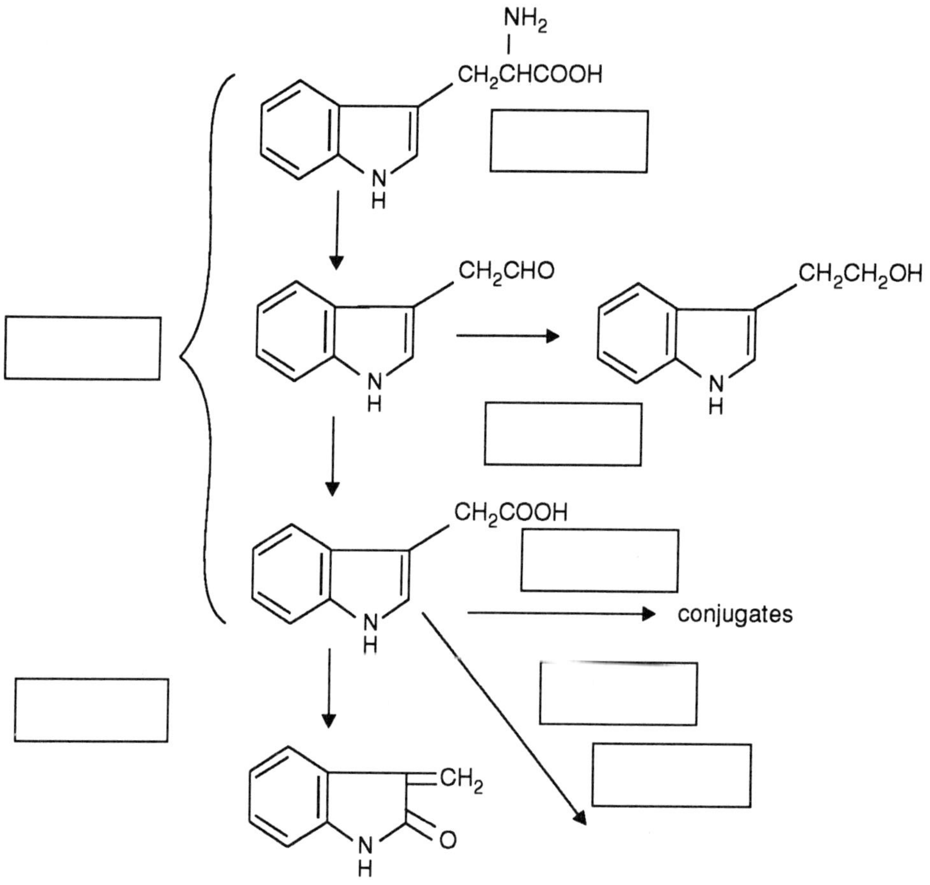

6.6.3 Synthetic auxins

IAA, in addition to being oxidised by IAA oxidase, is photolabile. It has to be kept away from bright light especially when in solution, and so extractions to determine tissue content must be carried out in dim light. A large number of compounds are known which are active in auxin bioassays but which are stable in bright light and not degraded by IAA oxidase. These compounds are not naturally-occurring but have been synthesised by organic chemists. They are called synthetic auxins and examples are shown in Figure 6.14.

We noted in Figure 6.3 that whereas low concentrations of IAA stimulate growth, high concentrations are inhibitory. At even higher concentrations it is lethal: plants are killed by auxin overdose. However, the concentration of auxin needed to kill narrow-leaved plants, for example cereals, is 10 times that needed to kill broad-leaved plants, such as poppy. This permits the use of some synthetic auxins as selective weedkillers or herbicides.

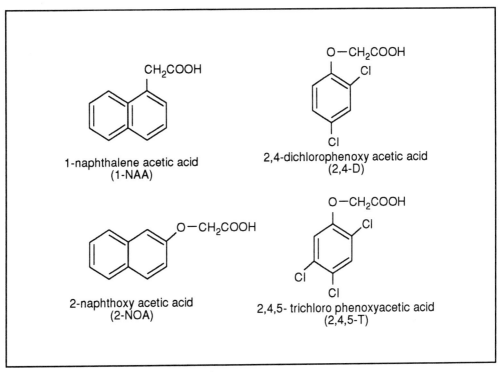

Figure 6.14 Examples of some synthetic auxins.

SAQ 6.5   Why would you choose a synthetic auxin rather than IAA as a herbicide?

Herbicides are commercially very important and literally hundreds of synthetic auxins have been synthesised. Details of most of these structures are, however, locked away deep in the recesses of chemical companies' vaults!

6.6.4 Auxin receptors and anti-auxins

Many of the compounds synthesised by companies searching for the perfect auxin were discarded because of lack of biological activity. A small number, however, interfered with the action of auxin and these are called anti-auxins (Figure 6.15). (Some more abbreviations for you to learn!).

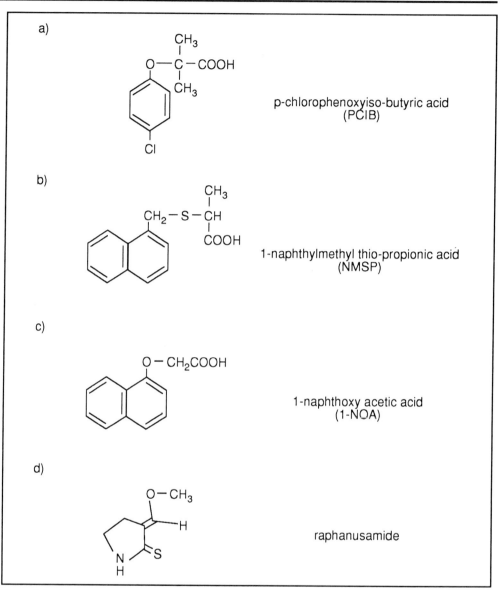

Figure 6.15 Structures of some anti-auxins; a), b) and c) are synthetic, d) is naturally-occurring; further details of this newly discovered group of compounds can be found in a paper by Sakoda *et al* Phytochemistry in Vol. 32 pp 1371-1373 (1993).

Analysis has shown that PCIB and NMSP do not reduce IAA biosynthesis. An alternative mode of action must therefore be sought and this allows us to introduce the topic of hormone receptors. If an auxin is to modify a cell's metabolism in some way it must first bind to a receptor thus generating a signal to which the cell responds. All hormone receptors are thus considered to be defined as molecules containing precise recognition sites to which the hormone binds. If you think for a moment you will realise that hormone receptors are just as important as hormones themselves and much effort has been applied to their study. They have proved to be difficult to isolate and characterise but a series of rules has evolved which allow us to predict that binding:

• will be reversible, since reactions should stop when the hormone is removed;

- will be of high affinity, because endogenous hormone concentrations are low;

- will be maximal at concentrations similar to those found in tissues;

- will be specific to active hormones only (but see below);

- will be confined to tissues which react with the hormone;

- will be linked to a biological response.

Anything which prevents the formation of the hormone-receptor complex will reduce the activity of the hormone and the anti-auxins are considered to do this. It is proposed that anti-auxins are sufficiently similar to auxins to compete with them for binding to the auxin receptor but sufficiently different so that the hormone: anti-auxin complex does not activate the cell's metabolism. Assuming that the formation of this inactive complex is reversible the level of activity evoked will depend on the relative concentrations of the auxin, the anti-auxin and the receptor.

SAQ 6.6

Experiments were carried out to determine the effect of 1-NAA and 2-NAA on the growth of tomato roots in a culture medium. The results shown below are for growth in length after seven days and are means of 10 replicates.

Experiment 1	Length (mm)
1) Control	128.8
2) 10^{-5} mol 1^{-1} 1-NAA	100.2
3) 5×10^{-5} mol 1^{-1} 1-NAA	93.2
4) 10^{-4} mol 1^{-1} 1-NAA	63.2

Experiment 2	Length (mm)
1) Control	133.0
2) 10^{-5} mol l^{-1} 2-NAA	115.6
3) 10^{-5} mol l^{-1} 2-NAA + 10^{-5} mol l^{-1} 1-NAA	124.7
4) 10^{-5} mol l^{-1} 2-NAA + 5×10^{-5} mol l^{-1} 1-NAA	134.0

Look again at Figures 6.3, 6.14 and 6.15 and suggest an interpretation for these results (data based on Weston, Canadian Journal of Botany, 1970, *48*, 2193-2197).

oligosaccharins

All but one of the anti-auxins shown in Figure 6.15 are synthetic compounds but there is now clear evidence for other naturally-occurring anti-auxins which may be involved in the control of cell elongation. One such compound is an oligosaccharide containing xylose and glucose residues, called XG9. XG9 occurs as a break-down product of one of the components of the cell wall and is a member of a group of compounds called oligosaccharins. We will briefly discuss the broader role of such compounds later in the chapter.

Radishes and a number of other species are known to produce compounds which inhibit shoot extension growth and contribute to phototropic bending. These compounds have been identified in radish and the structure of one of them, raphanusamide, is shown in Figure 6.15. The compounds from radish all interfere with the action of IAA in the oat coleoptile bio-assay and are, therefore, anti-auxins.

raphanusamide

This discussion of anti-auxins concludes our examination of the factors affecting steady-state auxin concentration. We will now look at each of the remaining classes of hormone in turn, following the pattern we have established with auxins.

6.7 Gibberellins

Gibberellins (GAs) were discovered as a result of the study of the *bakanae* disease of rice, in which plants grow so tall that they cannot hold themselves upright. The disease is caused by the fungus *Gibberella fujikuroi*, which releases compounds into the tissue of the plant, these compounds being responsible for the excess growth. Several of these compounds were isolated and identified and subsequently shown to occur naturally in plants. They are structurally very similar to each other, having the gibbane carbon skeleton shown in Figure 6.16.

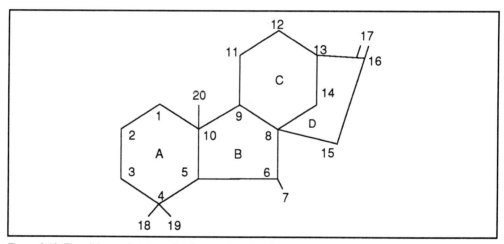

Figure 6.16 The gibbane skeleton, showing numbered carbon atoms.

Whilst being structurally very similar to each other, these compounds are obviously very different from auxins. Further, they show no activity in the oat coleoptile growth test. They were, therefore, classed as a new group of plant growth hormones and some of their actions on plants are listed in Table 6.6.

Add these items to your summary sheet of physiological/morphological effects of hormones.

Physiological and morphological effects of GAs

    i)   release of seed and bud dormancy

    ii)   hypocotyl and internode elongation

    iii) control of cambial activity

    iv) production of digestive enzymes in germinating cereals

---

Table 6.6 Some physiological and morphological effects in which GAs are involved.

6.7.1 Biosynthesis and degradation of GAs

Gibberellins are terpenoids (compounds made up of isoprene units).

Isoprene units have the following structure:

$$- CH_2 - \overset{\overset{\displaystyle CH_3}{\displaystyle |}}{C} = CH - CH_2 -$$

Isoprene units contain five carbons and are joined head to tail to make terpenoids. GAs are terpenoids which contain 20 carbons derived from four isoprene units. The biosynthetic pathway is shown in Figure 6.17.

You are not expected to memorise this pathway! It is included for reference and to enable a number of points to be made. The pathway begins with the phosphorylation of mevalonic acid (Step 1) to produce dimethylallyl pyrophosphate which is in equilibrium with isopentenyl pyrophosphate (Step 2). The pathway is characterised by the progressive addition of isoprene units such that a 10 carbon compound is produced in Step 3, a 15 carbon compound in Step 4 and a 20 carbon compound in Step 5. There are no further additions of an isoprene unit, for we have already reached the 20 carbons needed for a GA. The remainder of the pathway consists of three features; the cyclisation achieved in stages 6 and 7 to produce *ent*- kaurene; the progressive oxidation at carbon 19, achieved in Steps 8,9 and 10; and finally the hydroxylation of carbon 7 accompanied by rearrangement to convert ring B into a 5-membered ring in Steps 11 and 12. The product of the pathway, GA_{12} aldehyde, is considered to be the precursor of all other GAs. Before we leave the pathway, note the branch point at farnesyl pyrophosphate, which leads to steroids and that at geranyl-geranyl pyrophosphate, which leads to carotenoids. It is important to realise that the early part of the pathway is shared by a number of important biosynthetic pathways and we will return to this point later.

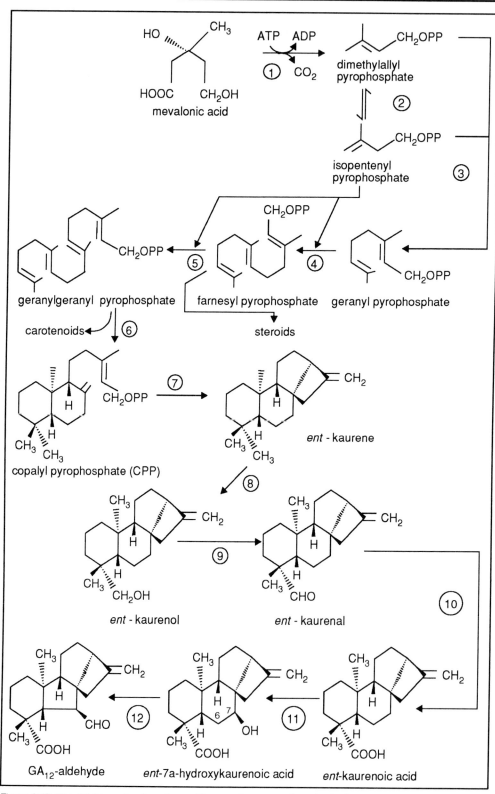

Figure 6.17 The GA biosynthesis pathway. Step numbers are referred to in the text.

bioassays of GAs

Altogether, more than 70 GAs of varying activity have been isolated from various organisms, 60 of which are found in higher plants. A single species, however, would not be expected to contain more that 15. How do we decide if some of the GAs are more active than others? The answer is their activities are assessed in bioassays. Three bioassays are used in particular; growth of the lettuce hypocotyl, growth of the stem in dwarf peas and growth of the leaf sheath in dwarf rice. These tests and a number of others allow us to draw up a series of rules governing the increase or decrease of GA activity.

1)   Activity increases when carbon 7 is oxidised to COOH.

2)   Activity increases as carbon 20 is oxidised from -CH₃ through -CH₂OH to -CHO and is finally lost to form 19-carbon GAs. In the 19-carbon GAs, carbon 19 forms a lactone ring over the A ring, linking carbon 10 with carbon 4 (see GA₁ in Figure 6.18 below).

3)   Activity increases with insertion of an -OH group at carbon 3 or carbon 13 and is highest with substitutions at both.

4)   Activity is severely lowered by the presence of an -OH group at carbon 2 and further reduced by the breaking of the lactone ring and the oxidation of carbon 19 to -COOH.

Examine the structures shown in Figure 6.18 and, using the information provided above, place them into a sequence from precursor through progressive activation to final loss of activity. If the solution to this question is visible to you, cover it up until you have tried to work it out.

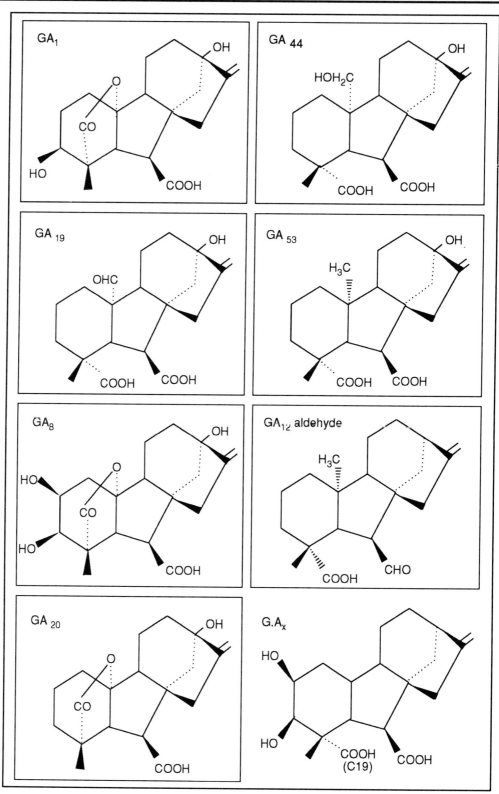

Figure 6.18   Structure of seven actual and one postulated GA.

The correct sequence is shown in Figure 6.20.

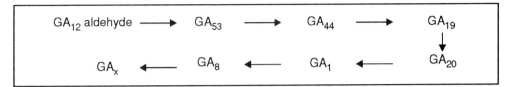

Figure 6.19 Biosynthetic pathway of GA₁ and its subsequent inactivation

∏ If you re-draw Figure 6.19 but put in the structures of the compounds used, you
will see how these compounds relate to each other structurally. The easiest way
to do this is to photocopy Figure 6.18 and cut out the structures.

Note that GA_x has never been found in organisms but a compound has been identified
with a ketone instead of -OH at carbon 2. This is given the name GA_8 metabolite.
Subsequent reactions in the degradation sequence break open the rings, thus removing
the basic GA structure.

6.7.2 Conjugation also occurs with GAs but not polar transport

GA conjugates are also formed but only with glucose. Conjugation either occurs at
carbon 2, forming a glucosyl ether, or at carbon 7, forming a glucose ester. Some species
can hydrolyse conjugates to the free GA and here the conjugate appears to be a storage
form. Other species lack this ability and the formation of the glucose ester of GA, for
example, would constitute inactivation. Thus, as with auxin conjugates, GA conjugates
do not appear to have a single unifying role.

GAs found in
xylem and
phloem

Experiments have been conducted to seek evidence for a polar transport system for
GAs, analogous to that for auxins, but none has been found. GAs are found in *both* xylem
and phloem sap but there appears to be no special mechanism regulating their
distribution.

6.7.3 Synthetic GAs and anti-GAs

Perhaps because of the complexity of their chemistry, no synthetic compounds with GA
activity are available. However, the compound GA_3, which differs from GA_1 only in
having a double bond between carbon 1 and 2, is available in plentiful supply in culture
filtrates of *Gibberella fujikuroi*. Although GA_3 is not a major GA in plants, plants do have
the ability to convert it to more active GAs. A mixture of two other GAs, GA_4 and GA_7,
is also available. Thus, we have the means to increase the GA content of a tissue by the
application of these compounds.

growth
retardants
(antigibberellins)

We also have the means to reduce the GA content of a tissue, by the use of anti-GAs,
sometimes called dwarfing agents or growth retardants. This group comprises a large
number of synthetic compounds which reduce internode elongation without obvious
phytotoxicity. In this they differ from herbicides which, of course, kill plants. Anti-GAs
have found considerable use in restricting the height of cereals, thereby preventing
lodging by wind and rain. They are also used to produce short stocky growth in
numerous ornamental species. The application of GAs overcomes their effect, hence the
name anti-GA.

Compounds with anti-GA activity have been available for many years but the original ones, for example, daminozide and chlorocholine chloride, have largely been replaced by a group of recently developed compounds, which are all derivatives of triazole. The triazole retardants are active at much lower concentration than previously-used compounds and their mode of action is much more clearly understood. They all inhibit oxidation Steps, 8, 9 and 10 shown in Figure 6.17 on the GA biosynthetic pathway and so reduce GA biosynthesis. The naturally-occurring compound catechin has recently been shown to inhibit the oxidation of GA_{12} aldehyde to GA_{53} and thus to have similar effects to the synthetic anti-GAs. Catechin is not widely distributed in plants, however, and appears to be restricted to the seed coat.

triazole retardants

catechin

SAQ 6.7

Answer the following questions.

1) Would you expect that a reduction in GA concentration would be the only effect brought about by application of a triazole retardant?

2) How do anti-GAs differ from anti-auxins, other than structurally?

3) Is the following statement true? Anti-auxins and anti-GAs are all synthetic compounds, they do not occur naturally.

We are now ready to proceed to the next hormone class.

6.8 Cytokinins

We will introduce cytokinins by examining the response of plants to wounding. When a plant is wounded, (eg wind damaged, grazing by animals) the plant responds as we do by producing new cells to fill in the wound. The wound tissue is called callus and it consists of parenchyma cells which are, as you remember non-specialised. Pieces of callus can be excised, cultured in a growth medium and their development studied.

SAQ 6.8

Bearing in mind that callus is non-photosynthetic, what must we provide in a growth medium in addition to water to keep the callus alive?

basal medium

In addition to minerals and sucrose we need to add small quantities of vitamins. Whole plants can manufacture all the required vitamins but when parts of plants are cultured, certain vitamins must be supplied in the growth medium. Such a medium would be described as basal medium (BM). When pieces of callus tissue are cultured in BM the cells remain alive but they grow and divide very little. Addition of an auxin, but not GA, brings about considerable cell enlargement but not cell division. A large number of compounds were tested for their ability to cause cell division when added to BM with auxin, and the first success was obtained in 1956 when herring-sperm DNA was used.

kinetin and cytokinins

Note that plant tissue culture medium, like microbiological medium, must be sterilised before use, and the simplest way to do this is to autoclave it. Autoclaving DNA breaks it down and the compound 6-furfurylamino purine is among the products. This compound was shown to be responsible for the cell division activity of the herring-sperm DNA. It is given the name kinetin but, despite its formation from DNA, does *not* occur naturally. Several compounds with similar activity have, however, been isolated from plants. They are called cytokinins (CKs) and the structures of some of them are illustrated in Figure 6.20.

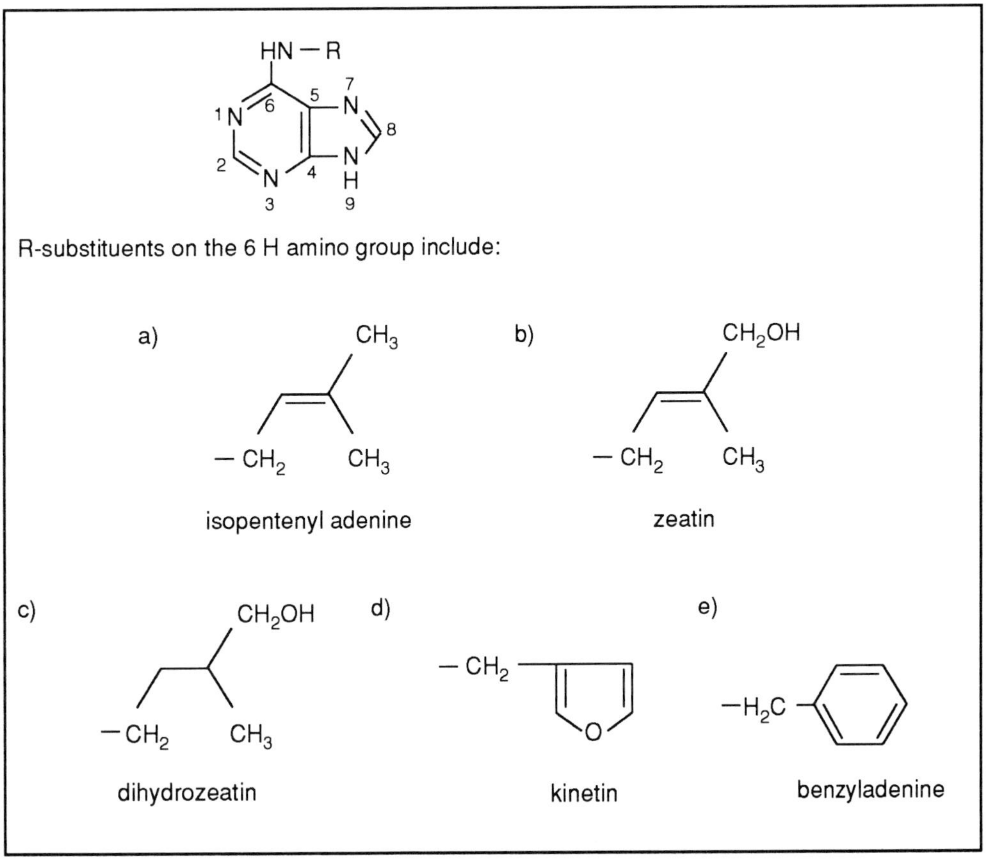

Figure 6.20 Structures and names of endogenous [a), b) and c)] and synthetic [d) and e)] cytokinins.

SAQ 6.9

Testing the effect of herring-sperm DNA as a plant cell division factor may seem bizarre but, in fact, it was quite logical. Why?

CKs are similar to auxins and GAs in that they are involved in a number of developmental processes and some of these are shown in Table 6.7.

Physiological/morphological effects of cytokinins

    i)   retardation of leaf senescence

    ii) stimulation of axillary shoot growth

    iii) stimulation of seed germination

    iv) stimulation of cell division and shoot differentiation

    v) stimulation of cell enlargement in leaves

Table 6.7 Some physiological effects of CKs.

⊓⊔ Add the list of effects of cytokinins to your summary sheet of the physiological/morphological effects of hormones.

6.8.1 CK biosynthesis

With one exception that we will discuss later, all natural CKs are adenine derivatives. In addition to occurring as the free base, CKs are also found as the riboside (a ribose sugar is attached at position 9 on the ring) and as the ribotide (the ribose sugar moiety is esterified with phosphoric acid at its 5 position). In fact, ribotides and ribosides precede the formation of the free base, during CK biosynthesis. The biosynthetic scheme is shown in Figure 6.21.

The key enzyme in CK biosynthesis is cytokinin synthase, which catalyses the addition of an isopentenyl group to AMP to form isopentenyl adenosine monophosphate. Figure 6.21 shows the proposed pathway for conversion of this into the riboside and free base, on the one hand, and the zeatin family on the other. The dihydrozeatin family is considered to be produced by reduction of zeatin but it is not known at what stage this occurs. The question arises as to whether the free base, riboside or ribotide is the active form of CK but no definite answer is available despite considerable effort by many research groups.

bioassays of CKs

CK-specific enzymes

As with auxins and GAs, bioassays have been used to determine the activity of compounds which we think have activity as CKs. Numerous bioassays are available for CKs but the three most widely used are the prevention of loss of chlorophyll in excised cereal leaves, the synthesis of betacyanin by *Amaranthus caudatus* and the stimulation of cell division in callus cultures. Numerous species have been used in the callus bioassay but the soybean test appears to be the best because no substances other than CKs are known to be active. For this reason soybean callus is a very useful material for metabolic studies. These studies show that the free base, nucleoside and the nucleotide can readily be absorbed by tissues and quickly converted into the other forms. This suggests that plants contain CK-specific enzymes that catalyse base-nucleoside-nucleotide inter-conversions and the activity of these enzymes makes it difficult to decide which is the active form of CK.

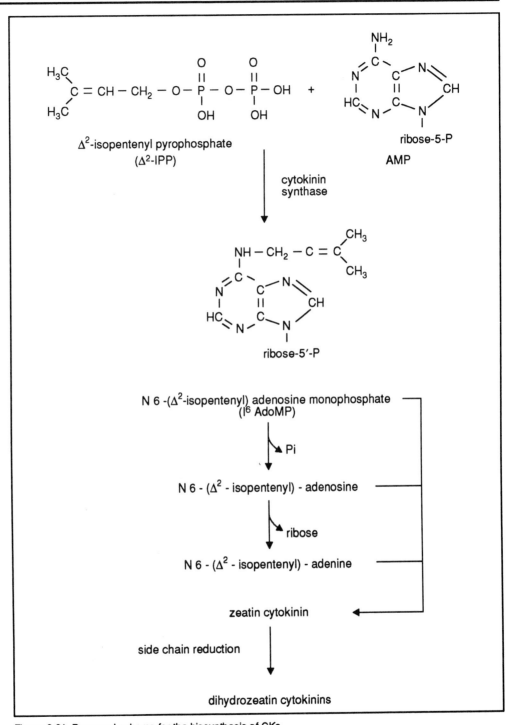

Figure 6.21 Proposed scheme for the biosynthesis of CKs.

coconut milk
as a source of
CKs

Prior to the discovery of CKs the liquid endosperm of coconut, called coconut milk , was often added as a growth supplement because of its ability to support cell division. (Figure 7.7d shows a diagram of the coconut and the location of the milk) Analysis of coconut milk subsequently showed the presence of adenine CKs. However, cell division-promoting activity was also found associated with the compound diphenyl urea. Since then numerous substituted phenyl-ureas have been synthesised and shown to be active in CK bioassays (Figure 6.22).

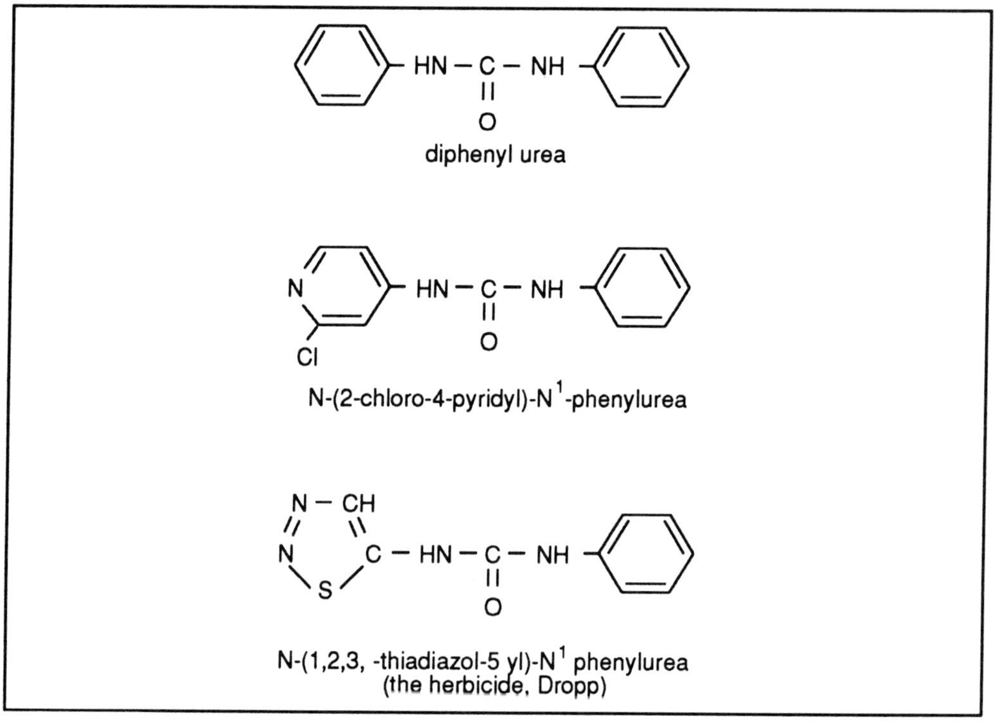

Figure 6.22 Structure of phenyl-urea cytokinins.

This was puzzling, since the diphenyl ureas and the adenine CKs show no structural similarities. A recent discovery offers a solution. Removal of the five-carbon side-chain of CKs causes the virtual total loss of CK activity and this cleavage is catalysed by an enzyme called cytokinin oxidase. This appears to be the major degradation pathway for cytokinins. It has recently been shown that the phenyl ureas inhibit the *in vitro* activity of CK oxidase.

SAQ 6.10	If a diphenyl urea inhibits CK oxidase activity *in vivo* what effect should this have on the CK level of a tissue?

Diphenyl urea application does, indeed, lead to increased tissue CK content which is in agreement with an action on CK oxidase.

Before we leave this section we should mention the occurrence of CK in transfer RNA (tRNA). Some types of tRNA contain adenine with an isopentenyl adenosine. When these tRNA molecules are degraded, isopentenyl adenosine is released and the question arises as to what contribution this makes to total CK levels. The general consensus is that it is negligible. Soybean callus has an absolute requirement for CK ie CK must be added for cell division to occur, despite the fact that it contains several tRNA species which contain isopentenyl adenosine. Also, the level of CKs in pea root tips has been found to be 27 times that present in tRNA. Thus, CK from tRNA is not thought to make a significant contribution to free CK levels.

6.8.2 Conjugation and transport

conjugates
with glucose
and alanine

Several conjugates are found between CK and glucose and one with alanine. Conjugates with glucose are formed at the 3-, 7- or 9- positions of the purine ring and through the side chain -OH of the free base. The conjugate with alanine occurs at position 9 of the ring. The 3-glucoside and side chain glucoside are as active as free CKs but the others show virtually no activity. Thus, the 3- and side chain glucosides can be looked upon as storage forms with the others as inactivation products. As with auxins and GAs, different species vary with regard to which of these conjugates they produce.

CKs produced
in root
meristems

Root apical meristems are major sites of synthesis of CKs in whole plants and they are transported to the shoot in the transpiration stream. They are also found in high concentration in shoot apices and in developing seeds, but there is disagreement as to whether they are synthesised in shoot apices and seeds or simply transported there. The enzyme cytokinin synthetase has recently been discovered and studies on its distribution should throw light on this situation. CKs appear to be part of the plant's root to shoot signalling system because environmental factors that interfere with root function, such as water stress, reduce the CK concentration of xylem exudate. We noted above that CKs retard leaf senescence. This can be demonstrated by the application of discrete drops of a CK onto the surface of excised leaves. In such cases the treated area remains dark green while the surrounding tissues become yellow. This implies that the CK has not been translocated away from the point of application. CKs have not been reported in phloem sap so, apart from movement in the transpiration stream, they are relatively immobile.

6.8.3 Synthetic and anti-CKs

first synthetic
CKs, kinetin
and benzyl
adenine

The first CK discovered, kinetin (Figure 6.20), is a synthetic CK and although formed from DNA, has never been found occurring naturally in tissues. Benzyl adenine was discovered soon after kinetin. The synthetic CKs are much cheaper than the natural ones and are widely used in plant tissue culture media. They are susceptible to CK oxidase activity, which contrasts with synthetic auxins, which are resistant to IAA oxidase activity. A considerable amount of work has been done in the area of anti-CKs and several compounds have been produced with this activity (Figure 6.23).

The two compounds in Figure 6.23, with a suitable side chain as the R group, inhibit cell division in callus cultures and CKs counteract this. Note their structural similarity to the adenine cytokinins. They are considered to compete with natural CKs for a receptor site and are, therefore similar in action to the anti-auxins. Compounds are also available which antagonise diphenyl urea activity. Thus, we are able to manipulate CK action both by the application of synthetic CKs and of anti-CKs. Whilst synthetic CKs are relatively cheap, the anti-CKs are very expensive. This means that they are available only for laboratory, but not field, experiments.

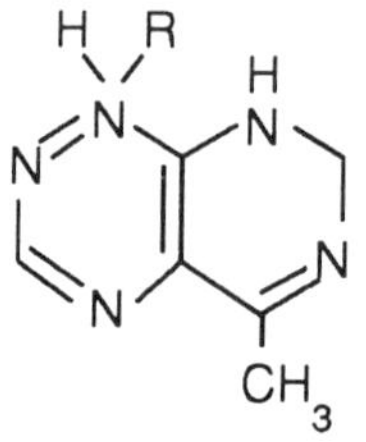

a pyrazolo [4,3-d] pyrimidine

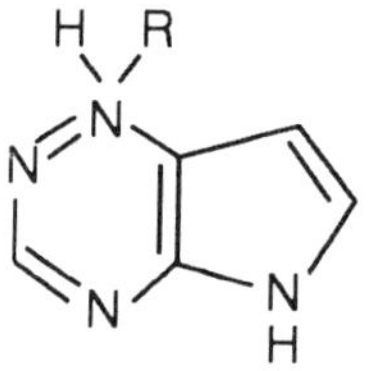

a pyrollo [2,3-d] pyrimidine

Figure 6.23 Structure of two anti-CKs.

We will now turn our attention to the next group of plant hormones, the abscisins.

6.9 Abscisins

This group of compounds was discovered simultaneously in two laboratories. Wareing's group was studying the dormancy of birch buds and isolated a compound they called dormin. Cornforth's group was investigating leaf and fruit abscission in cotton and they named their compound abscisin. Subsequent analysis showed the two compounds to be identical and the name abscisic acid (ABA) was decided upon. You may remember that we have met this compound before.

Π   Try to remember where we met ABA before. What other plant hormone appears to share the property of ABA that we have already met?

We met ABA in the water relations chapter; it is used to provide information to the shoot from the root. CKs are also involved in this process.

The structure of abscisic acid (ABA) is shown below:

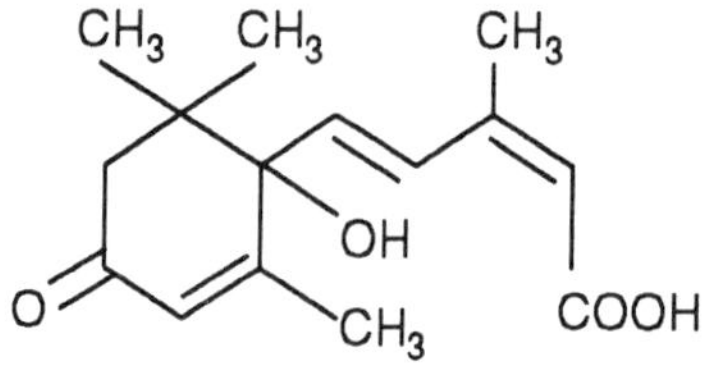

(+) 2-cis ABA

ABA exists in both *cis* and *trans* configurations, only the *cis* form being biologically active. Light causes the isomerisation of the *cis* to the *trans* form and this is considered to occur during its extraction from the plant. Some of the effects of ABA are shown in Table 6.8.

Physiological and morphological effect of abscisic acid

   i)   promotes leaf senescence

   ii)   inhibits growth of oat coleoptiles

   iii)   inhibits production of amylase by cereals seeds

   iv)   controls seed dormancy in some species

   v)   involved in the control of root geotropism

   vi)   involved in the control of stomatal aperture

Table 6.8 Some effects of ABA.

∏ Add the information given in Table 6.8 to your summary sheet.

6.9.1 Biosynthesis and metabolism

ABA is composed of three isoprene units and its biosynthesis involves the terpenoid pathway. Two routes are proposed for its biosynthesis, a direct route from mevalonic acid via farnesyl PP and an indirect route by the oxidative cleavage of a C_{40} carotenoid, violaxanthin, to yield a 15-carbon intermediate which is then converted to ABA (Figure 6.24).

evidence for the indirect synthesis of ABA

Although radio-tracer studies show that carbons from mevalonic acid are incorporated into ABA in the way predicted by the direct pathway, the amount of incorporation is very low indeed. Favour has recently swung to the indirect pathway. In the indirect pathway, oxygen is added to violaxanthin, as shown in Figure 6.24. Using wilting leaves which synthesise ABA very quickly, it can be shown that, in the presence of $^{18}O_2$, ABA is synthesised carrying the labelled oxygen in its carboxylic acid. This is where it would occur if ABA was produced by the oxidative cleavage of violaxanthin, but not by the direct pathway. Significant quantities of the 10-carbon by product can also be detected. Further, seedlings grown in the presence of the herbicides fluridone and norflurazon, which inhibit carotenoid biosynthesis, contain greatly reduced ABA levels.

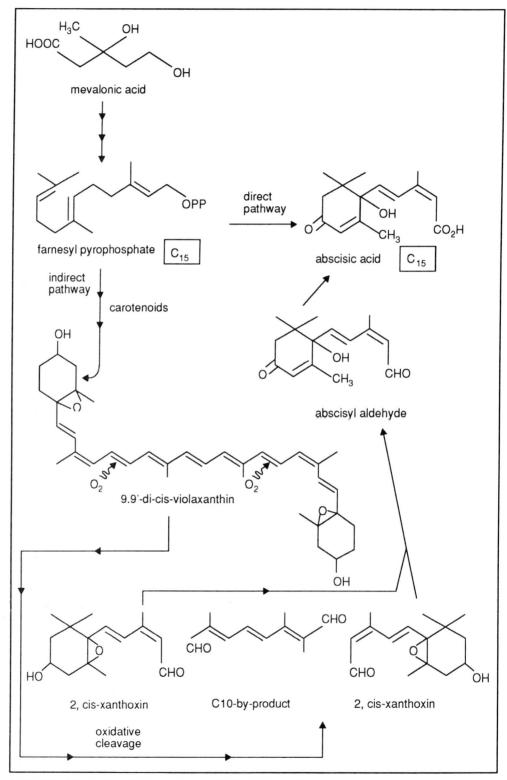

Figure 6.24 The direct and indirect pathways of ABA biosynthesis.

ABA can be readily metabolised to produce phaseic acid and the two forms of dihydrophaseic acid (Figure 6.25). These compounds show low or no activity in bioassays and their production, therefore, constitutes inactivation. ABA also forms conjugates with glucose; the most prevalent is ABAGE, the glucose ester formed via the carboxyl group but another glucoside, ABAGS can also be formed via the hydroxyl group on the 1' position. These conjugates do not appear to be a ready source of free ABA.

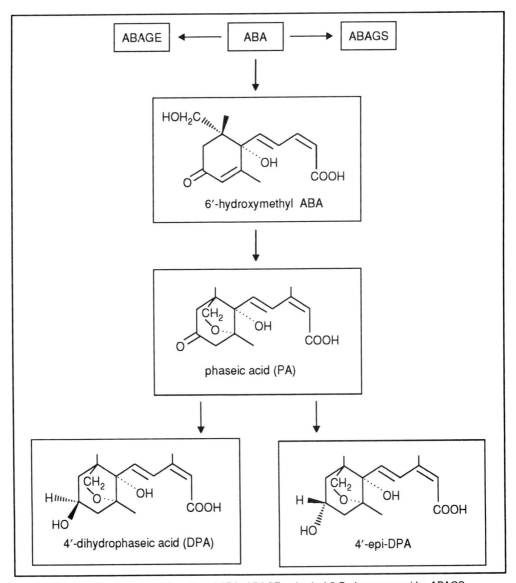

Figure 6.25 Degradation and conjugation of ABA. ABAGE, -abscisyl-β-D-glucopyranoside. ABAGS, 1-O-abscisic acid-β-D-glucopyranoside.

ABA is synthesised in both roots and shoots. We have seen already that it is transported in the xylem; it has also been found in phloem sap where its transport is prevented by steam ringing.

fluridone and
norflurazon are
anti-abscisins

We noted above that the two herbicides fluridone and norflurazon cause a reduction in ABA concentration by inhibiting a step in the production of the precursor carotenoids. This is analogous to the action of anti-gibberellins. Thus these two compounds can be considered to be anti-abscisins. Synthetic abscisins are not readily available but ABA is relatively inexpensive so, when elevated abscisins are needed, ABA itself can be applied.

6.9.2 Abscisic acid and anti-transpirants

In addition to modifying water consumption, the control of stomatal aperture may also affect chilling injury and the uptake of pollutants. The ability to reduce stomatal aperture could, therefore, have a significant affect on plant growth and development. Because of its effects on stomata, a considerable effort has been made to ascertain the commercial potential of ABA application. ABA application does, indeed, reduce the transpiration rate of barley and coffee for example, but its effect lasts for only about 10 days. The application can be repeated, of course, but the costs of labour and materials renders this non-economical. Thus ABA is not presently in use as an anti-transpirant on a field scale.

6.10 Ethylene

The idea that volatile compounds could affect plant growth dates from the mid 1800s when observations were made on the ability of coal gas, burnt to provide illumination, to cause leaf abscission. Ethylene was subsequently isolated from coal gas and shown to be responsible for the effect on plants. Ethylene has since been shown to be a natural product of plants and some of its effects are shown in Table 6.9

Physiological and morphological effects of ethylene

i)   promotes leaf and fruit abscission

ii)   stimulates fruit ripening

iii)   stimulates germination in, for example, peanuts

iv)   stimulates flowering in Bromeliads such as pineapple

v)   inhibits root and shoot extension growth but stimulates radial growth

vi)   induces shoot diageotropic growth (see text)

Table 6.9   Some effects of ethylene on plants.

∏   Add the effects of ethylene listed in Table 6.8 to your summary sheet of the physiological and morphological effect of hormones. (You should now have a sheet combining the information given in Tables 6.2, 6.5, 6.6 and 6.7 and 6.8 - giving information about all of the hormone groups).

When a shoot or a root grows such that its axis lies at a right angle to the direction of gravity it is said to be diageotropic. Shoots normally grow away from gravity and are said to be negatively geotropic. Roots grow towards gravity and are positively geotropic. As noted above, ethylene causes shoots to grow horizontally. This and the effects on extension and radial growth noted in Table 6.9 are referred to as the 'triple effects of ethylene' and they have often been used as a diagnostic test for the gas.

6.10.1 Biosynthesis and degradation

It is now established that ethylene is generated from carbons three and four of methionine. The process involves a cycle which conserves the sulphur atom (Figure 6.26), permitting a steady rate of ethylene production even when methionine concentration is low.

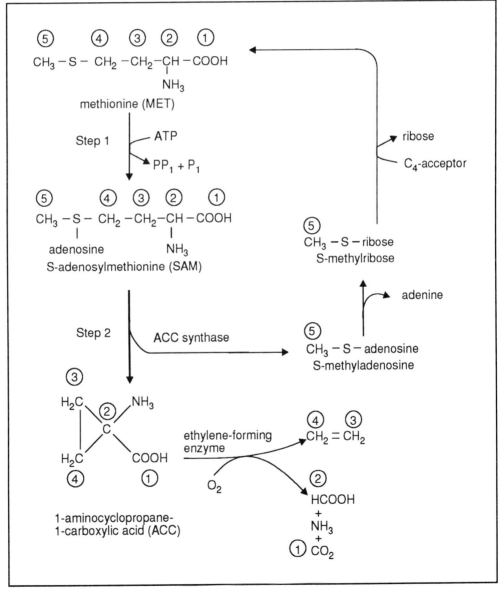

Figure 6.26 Ethylene biosynthesis. Note numbering of the carbon atoms.

ethylene
produced via 1
amino-
cyclopropane
1-carboxylic
acid

The cycle involves the reaction of methionine with ATP to produce S-adenosyl methionine. This is cleaved to form 1-amino-cyclopropane-1-carboxylic acid (ACC) and S-methyladenosine (SMA). SMA is rapidly hydrolysed to S-methyl ribose and then converted back to methionine. Aminocyclopropane-1-carboxylic acid (ACC) is broken down to produce ethylene, CO_2 and ammonia, the latter product becoming incorporated into asparagine.

The process is controlled mainly by two enzymes; ACC synthase, which produces ACC, (Step 2) and the ethylene-forming enzyme (EFE) which catalyses the breakdown of ACC (Step 3). Ethylene application causes the increased synthesis of ACC synthase while wounding, chilling injury and drought stress cause an increase in its activity. EFE activity depends on the availability of oxygen. In its absence no ethylene is formed and ACC accumulates. EFE activity is also reduced by CO_2 and by temperatures above 35°C. Since ethylene is a gas it diffuses out of the tissue in which it is formed and this contributes to the regulation of its level in the tissue.

Ethylene itself is degraded by an enzyme complex termed ethylene oxidase, which inactivates it by catalysing the formation of ethylene oxide and ethylene glycol.

ethylene
effects in
waterlogged
soils

Oxygen is deficient in waterlogged soils and results in the accumulation of ACC in the roots. This is transported to the shoot. Under the influence of O_2 available at the base of the stem, exposed to air, the ACC is converted to ethylene. One of the actions of ethylene is to reduce the polar transport of IAA, which , therefore, accumulates towards the base of the stem, and causes the initiation of adventitious roots. These grow down into the surface layers of the soil where O_2 levels allow for normal ethylene biosynthesis. The phenomenon just described allows the plant, at least partly, to overcome the effects of waterlogging.

6.10.2 Conjugation and transport

inactivation of
ACC by
conjugation
with malonic
acid

No conjugation products are known using ethylene itself but ACC conjugates with the organic acid malonic acid to form N-malonyl ACC (MACC). MACC is a very stable product and does not break down to release ethylene. Further, MACC has no action on fruit ripening or any of the other processes stimulated by ethylene so its formation constitutes inactivation. It is assumed that once ethylene has been formed its distribution cannot be controlled by the plant. However, there is a transport element because ACC itself is translocated. Under waterlogging conditions considerable quantities are transported in the xylem to the shoot where a burst of ethylene evolution occurs. Most tissues appear to be able to synthesise ACC however, and its transport does not appear to be a significant phenomenon in well-aerated plants.

6.10.3 Synthetic and anti-ethylene compounds

If one wishes to increase the ethylene content of a tissue it is possible to expose it to the gas but this is impractical unless the tissues are contained within a closed space. Of greater use is the compound 2-chloro-ethyl phosphonic acid (ethephon).

$$Cl-CH_2-CH_2-PO_3^{2-}$$

ethephon as a
source of
ethylene

Ethephon is stable at pH values below 4 but above this pH it breaks down to release ethylene, chlorine and phosphate. The pH of plant cells is above 4, so the ethylene concentration of a tissue can be increased by treating it with ethephon. The concentrations of ethephon needed for ethylene activity are so low that the amounts of the chlorine and phosphate produced as by-products are negligible.

AVG and AOA
as inhibitors of
ethylene
production
Ethephon is very cheap and it is one of the most widely used PGRs. Ethephon use, therefore, allows us to increase the ethylene content of a tissue. There are, however, also ways to reduce it. Two compounds have been discovered which inhibit ACC synthase, aminoethoxy vinyl glycine (AVG) and amino-oxyacetic acid (AOA) (Figure 6.27). Unfortunately, these compounds are too expensive for large scale field-trials and their use is restricted to laboratory experiments. Treatment with cobalt ions also results in reduced ethylene production but the enzyme affected here is EFE.

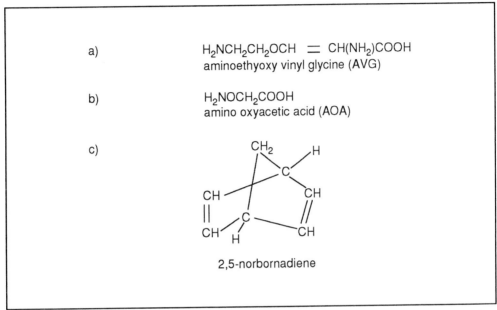

Figure 6.27 Structure of some compounds which affect ethylene metabolism. Compounds a) and b) inhibit ethylene synthesis; c) inhibits ethylene action.

Finally two compounds are available which inhibit ethylene action; silver ions and 2,5-norbornadiene (Figure 6.27). Silver thiosulphate is a highly mobile form of silver and is routinely used to retard senescence and extend the shelf-life of cut carnation flowers.

This brings us to the end of our systematic coverage of the five classes of hormones. By way of summarising aspects of this, carry out the following in-text activity.

∏ Fill in the boxes in Table 6.10a) with +, to signify that the hormone displays the property listed or - to signify that the hormone does not display the characteristic. (Cover up Table 6.10b which gives you the answers).

Property	Hormone				
	Auxin	GA	CK	ABA	Ethylene
exhibits polar transport					
anti-hormone, inhibits synthesis					
anti-hormone, inhibits activity					
forms conjugates with:					
a) sugars					
b) amino acids					
c) organic acid					
synthetic forms available					

Table 6.10a Comparison of properties of hormones.

Property	Hormone				
	Auxin	GA	CK	ABA	Ethylene
exhibits polar transport	+	-	- (?)	-	-
anti-hormone, inhibits synthesis	-	+	-	+	+
anti-hormone, inhibits activity	+	-	+	-	+
forms conjugates with:					
a) sugars	+	+	+	+	-
b) amino acids	+	-	+[1]	-	-
c) organic acid	-	-	-	-	+
synthetic forms available	+	-	+	-	+[2]

Table 6.10b Comparison of properties of homones. 1) CKs only conjugate with alanine. 2) If we count ethephon as a synthetic form of ethylene.

6.11 Some hormones affect the metabolism of others

In Chapter 8, we will see that hormones interact in their control of cell growth and differentiation but it is important to realise that they often interact with each other's metabolism. The best studied example of this is the effect of IAA on ethylene biosynthesis. IAA stimulates the synthesis of ACC synthase and, therefore, ethylene biosynthesis. Auxin treatment causes pineapples to flower but this can be shown to be the result of ethylene biosynthesis. In some systems it can be shown that the effects of supra-optimal amounts of auxin are caused by enhanced ethylene production. In cell cultures which require an exogenous supply of auxins, cytokinins are often found to have a 'sparing' effect, ie less auxin is required when a cytokinin is also present. This has been shown to be due to a lowered IAA oxidase activity, but the mechanism has not been elucidated. There are several other examples where the application of one hormone results in enhanced levels of other hormones and we are gradually elucidating the ways in which this occurs.

6.12 Mutants are useful in the study of hormones

Mutations in chromosomes and genes occur spontaneously and thereby generate new genomes. The rate of mutation can be increased considerably by artificial means, by the use of X-rays or other ionising (UV) radiation or by the use of mutagenic chemicals. Most mutations produce recessive traits and this will only be expressed when a plant is homozygous for the recessive character. Any mutation which produces a change which is lethal will lead to the death of the plant and will not be observed in the population. A considerable number of plant hormone mutants are now available, and we will examine GA mutants to illustrate their value.

some dwarf varieties are GA mutants

The existence of GA mutants has been known for many years. Gardeners know that they can buy seed of species such as peas and beans in either tall or dwarf varieties. The tall varieties produce long internodes and tall plants which need to be supported. The dwarf varieties, sometimes called bush varieties, produce short internodes, grow no more than 45cm high and do not need to be supported. These dwarf varieties are the result of natural mutations. In many cases their lack of height can be overcome by the application of GAs, so they are called GA mutants. Examination of these mutants shows that many of them are not able to synthesise GAs because of the lack of one or other of the enzymes which function on the biosynthetic pathway. Thus, one of the dwarf mutants is deficient in *ent*-kaurene synthetase whilst others lack one or other of the three oxidative enzymes which catalyse the conversion of kaurene to GA_{12} aldehyde.

mutants allowed identification of GA biosynthetic pathway

These mutants produce very low levels of GAs rather than none at all because many mutants tend to be *leaky*, that is, they sometimes produce low levels of an enzyme. Intermediates of the biosynthetic pathway accumulate in these mutants and their analysis and identification help us to propose a sequence for the pathway. The maize mutant called *dwarf*-1 is also a GA mutant and we now know that it lacks an enzyme prior to GA_1 production. This mutant can, therefore, be used to ascertain the biological activity of GAs which are produced before GA_1 in the biosynthetic pathway. These experiments show that GA_{53} and GA_{20} can stimulate the elongation of *dwarf*-1 plants, but GA_1 is a hundred times more effective. Finally at least one dwarf mutant is known which does not lack GA_1 or react to its application. Whereas there are numerous potential explanations for this, it is possible that this mutant is a GA response mutant, *ie* it does not respond to GA_1 perhaps due to an inability to bind the GA to a receptor site, which is necessary for activity. Thus, this mutant might possibly enable us to gain information about GA binding proteins.

| SAQ 6.11 |

Whereas mutants have been found which are concerned with various aspects of the metabolism of GAs, ABA and ethylene, no IAA biosynthesis or CK biosynthesis mutations are known. Provide a probable explanation for this.

Experiments were carried out to determine the ability of GAs and precursors to cause stem extension in maize dwarf mutants with the results shown in Table 6.11.

Answer the following questions (refer to Figure 6.19 if you need to).

1) How can the data be explained with regard to the biosynthetic pathway of GAs?

2) How would you change your answer to question 1) if GA_{19} produced a response of 100% in all mutants?

	GAs and GA precursor addition				
Mutant	GA_{12}	GA_{53}	GA_{19}	GA_{20}	GA_1
dwarf-x	0	95	98	98	100
dwarf-y	0	0	98	100	100
dwarf-z	0	0	0	98	100
dwarf-1	0	1	1	1	100

Table 6.11 Relative activities of GAs and GA precursors in bio-assays using maize dwarf-1, dwarf-x, dwarf-y and dwarf-z mutants. Data normalised with respect to the activity with GA_1.

6.13 Numerous other naturally-occurring compounds affect plant growth

In the chapter so far we have examined the five classes of hormone. There are numerous other groups of compounds which can modify plant growth and we will examine three of them briefly.

6.13.1 Oligosaccharins

oligo-saccharins can act as anti-auxins or mimic auxin activity

An oligosaccharin is a biologically-active, soluble oligosaccharide produced by the partial degradation of a cell-wall-bound carbohydrate polymer. We have already come across one example of an oligosaccharin, the xyloglucan-digestion-product XG9 which acts as an anti-auxin. A similar molecule but with seven xylose/glucose residues instead of nine mimics the action of IAA in stem growth tests. A related but separate group of oligosaccharins is derived from the 1-4-linked polygalacturonic acid which is a major constituent of the pectin fraction of cell walls. Oligomers from this source containing between four and eight sugar residues act in the same way as XG9 ie as an anti-auxin, whereas compounds with 12-14 residues can very strongly interact with IAA and CKs in stimulating root and shoot formation in tissue cultures.

One other example should be mentioned, even though it concerns an oligosaccharin of fungal origin. Fungal cell walls contain β 1-3 and β 1-6 glucan oligomers and these compounds cause plant cells to release phytoalexins, which are low molecular mass, anti-microbial compounds. It had been thought for many years that oligo- and polysaccharides were somewhat imprecise in their structure and that this would rule out their use as signalling agents. This has been shown not to be the case and now that methods have been improved for their synthesis, oligosaccharins are receiving increasing amounts of attention.

6.13.2 Polyamines

The occurrence of the polyamines, spermine, spermidine and putrescine (Figure 6.28) in plant tissues has been known for many years but only recently has the possibility been revealed that the actions of plant hormones might be mediated through changes in polyamine metabolism.

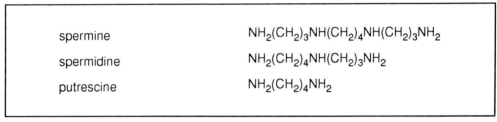

Figure 6.28 Structure of spermine, spermidine and putrescine.

Polyamines may well be involved with the action of all the hormones. However, they appear to have a closer relationship with ethylene because the compound S-adenosyl methionine occurs in the biosynthetic pathway of both groups. These pathways appear to compete with each other for this compound since the prevention of its use by one pathway leads to increased flux through the other. For example, if we inhibit ethylene production, the synthesis of polyamines increases. We are not able to decide at the moment whether polyamines should be considered to be secondary messengers for the major hormones or another class of hormones in their own right.

6.13.3 Brassinosteroids

brassin and
brassinosteroids

A number of plant scientists continue to seek new natural constituents with plant growth regulating activity and the group of compounds called brassinosteroids was discovered in the early 1970s as a result of such activity. Extracts of rape pollen (*Brassica napus* L.) caused a strong growth increase when applied to young pinto bean plants but the effects could not be explained by the presence of any of the established plant hormones. Initially the active component was called brassin but it was quickly shown that there were numerous active components and that they were steroid in nature. Thus, the name brassinosteroid (BR) was chosen as the parent name of the more than 30 components so far discovered. The structure of an important member, brassinolide, is shown in Figure 6.29. It is usually present at low concentration, for example 10 mg of brassinolide has been obtained from 200 kg of rape pollen.

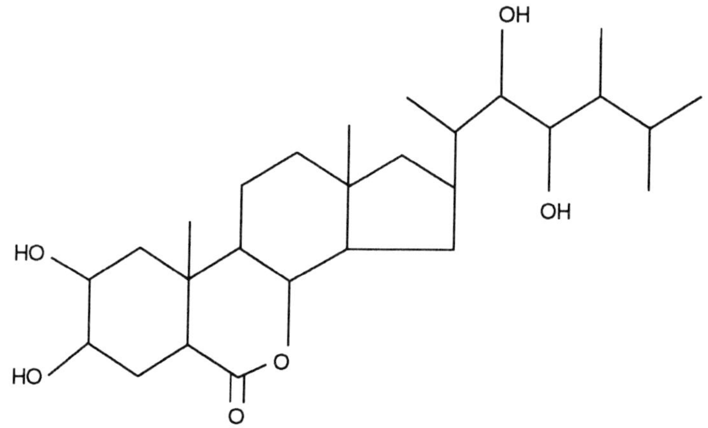

Figure 6.29 Structure of brassinolide, the first brassinosteroid to be identified.

Since their discovery in pollen, BRs have been detected in leaves, flowers, seeds, stems and roots in at least 22 species, thus they are not simply an oddity restricted to a single, somewhat unusual, location. However, their range of biological activities makes it difficult to classify them. In these activities they sometimes mimic the established hormones but sometimes interact synergistically with them. In no case, however, do they behave exclusively as if they belonged to one of the established groups. BRs tend to prolong the elongating phase of plant shoots and, in *Luffa cylindrica*, they caused the reversion of a floral to a vegetative apex. These and other pieces of evidence have led to the suggestion that BRs should be considered to be 'juvenile hormones', in some way responsible for retaining plants in their juvenile form. This and other possibilities are discussed more fully in an article by Mandava in Annual Reviews of Plant Physiology, 39, 23-52 (1988).

6.14 The mechanism of action of plant hormones

In this chapter, we have described the physiological/morphological effects of hormones but we have said little about the mechanisms by which the hormones exert their effects.

It is beyond the scope of this text to go into details yet it would be an omission to say nothing. It should not surprise you to learn that the complete position is not clear for any of the hormones. Possible mechanisms of action can be divided into two broad groups:

• a direct effect on the expression of genetic information;

• an effect on an aspect of intermediary biochemistry/physiology , ie not involving the expression of genetic information.

The evidence suggests that all five classes of hormone can act in both ways and some examples will be described in the final chapter when we look at aspects of growth and development.

Summary and objectives

In this chapter we have learnt that plants produce a number of compounds with attributes similar, but not identical, to those of animal hormones. Plant hormoness are sometimes called plant growth regulators (PGRs). However, the use of the term hormone for naturally-occurring growth active compounds and PGRs for synthetic ones does allow a distinction to be made and may avoid confusion The steady-state concentration of a hormone is important to the control of growth and can be determined by bio-assay, by immunological methods, by physico-chemical methods such as GLC, HPLC and MS or by combination of these. The factors which govern the steady-state concentration of a hormone are its rate of biosynthesis and rate of degradation, and its transport to or from a particular zone. In addition, hormones form conjugates with sugars and amino acids and these may act as storage forms or inactivation products.

We have provided details of the biosynthesis, degradation, transport and conjugation of auxins, GAs, ABA, CKs and ethylene. Information on their control is however, far from complete. We described that synthetic compounds are available with auxin, CK and ethylene action but not GA or ABA action. Anti-hormones are available for each class, which inhibit either the biosynthesis of the hormone or its action. Thus ways are available to increase or to decrease hormone concentrations or activity so providing potential for the manipulation of plant growth. Hormone mutants are also available for some hormones and these aid in the elucidation of the details of hormone biochemistry and physiology. We concluded the chapter by briefly describing oligosaccharins, polyamines and brassinosteroids which exhibit growth-active properties but which are not yet given the status of a class of hormone in their own right.

Now that you have completed this chapter you should be able to:

- define the term hormone, as used in animals and plants;

- show an understanding of how hormones are quantified by describing methods and analysing data;

- demonstrate a knowledge of factors affecting hormone steady-state concentrations and an ability to interpret supplied data relating to hormone concentration;

- show an understanding of ways in which hormone concentration or activity may be manipulated by chemical or other means;

- list the five classes of hormones and describe some of their effects on plants;

- give examples of compounds with anti-hormone activities.

Reproduction

Reproduction

7.1 Introduction

sexual and
asexual

In the previous chapters we have been following the theme that a plant's main aim is to survive and reproduce. We have examined a number of physiological processes that show how growth is in tune with the environment and are left with two major questions, namely how do plants grow and develop, and how do they propagate to ensure the survival of the species? We will discuss these questions out of order and deal with reproduction first. In the process of reproduction plants produce offspring in two fundamentally different ways termed sexual and asexual reproduction, the latter sometimes being referred to as vegetative reproduction. The important difference between these two processes is that in sexual reproduction genes from two parents are combined, producing progeny which differ from the parents. In asexual reproduction no recombination of genetic material occurs, so that the progeny contain the same genetic information as the parent and are termed clones. In this chapter, we will examine the two processes from a morphological and developmental point of view and then go on to discuss what we know about the control of sexual reproduction.

7.2 Asexual reproduction

The vegetative structures that plants use for asexual reproduction are modified roots and stems, and, in some cases, they also serve as a means of food storage.

7.2.1 Stem modifications

bulbs

Bulbs, for example onions, are shortened stems with thick fleshy leaves (Figure 7.1a). The stem extends to produce the flower and when flowering is complete new bulbs develop from the buds present in the axils of the leaves.

corms

Corms look similar to bulbs but are, in fact, quite different. A corm is shorter and broader than a bulb and consists of a swollen stem surrounded by thin scale leaves (Figure 7.1b). Flowers form from the buds at the top of the stem; corms often produce several flowering stems each. Each year a new corm forms at the base of each stem but, in addition, miniature corms called cormels, are produced between the new corm and the disintegrating old one. Examples here are gladiolus and crocus.

runners

Runners are specialised stems which develop from axillary buds by growing horizontally rather than vertically (Figure 7.1c). At every other node on the runner an axillary bud produces a cluster of leaves on a short stem. Roots develop at the base, producing a complete plant which is still attached to the parent. Strawberry is a common example and this process allows a single plant to spread and cover a wide area in a short period of time. Ordinarily the daughter plants remain attached to the parent but they can be easily separated by commercial growers who use this method for propagation. Runners are interesting physiologically because they produce long internodes whereas the parent and the progeny produce very short ones.

rhizomes

Whereas runners are horizontally-growing aerial stems, horizontally-growing underground stems also occur. These are called rhizomes. Rhizomes can be short and

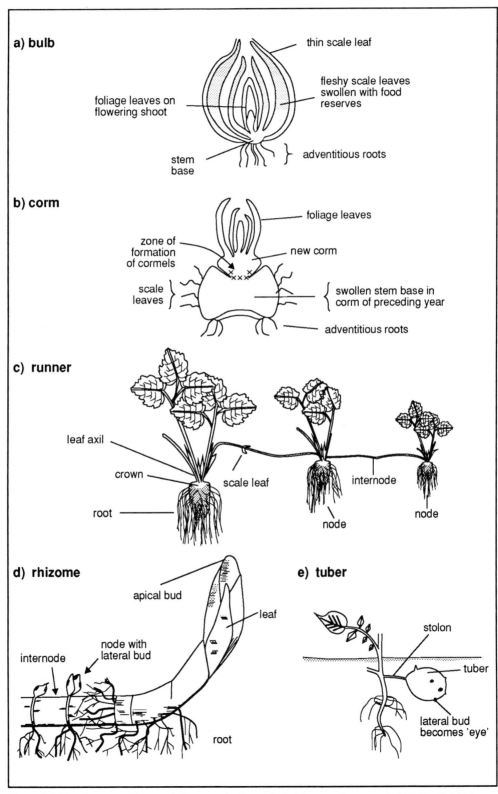

Figure 7.1 Structures used for vegetative reproduction a) bulb, b) corm, c) runner, d) rhizome, e) tuber.

fat or long and thin but they function in the same way as runners. New shoots are produced at nodes and these root easily to form whole plants. In iris the rhizome is thick and fleshy and may grow up to 20 cm in a year (Figure 7.1c). In twitch (couch grass, *Elymus repens*) the rhizome is long and thin and can grow several metres each year under good conditions. This method of spreading is used by a number of weeds which, because of this growth form, are particularly difficult to eradicate. Hoeing simply divides the original plant into a number of smaller ones which continue to grow.

stolons   The last example here is the special type of underground stems produced in potatoes, called stolons (Figure 7.1d). Each stolon is capable of producing a single tuber which develops as a swelling of the end portion of the stolon. The swelling is caused by proliferation of central parenchyma cells of the stem and the laying down in them of starch. The lateral buds on the original stolon become the 'eyes' of the potato and the original growing point becomes the apical bud. Growth subsequently occurs at the apical and lateral buds, roots forming at each. Thus, each tuber can form several plants and, because each parent can produce several tubers, this is a very effective method of propagation.

7.2.2 Root modifications

tubers   Some roots also produce swollen food storage organs which are involved in reproduction. These are also called tubers or, more usually tuberous roots. Adventitious shoot buds are produced on them and when these start to grow, roots are formed at the base. The sweet potato (*Ipomoea* spp) is an example. The term adventitious refers to the fact that an organ has developed in a location which is not its usual one. Thus, roots produced on a stem are called adventitious roots because it is not the normal place for them to grow, and shoots produced on a root are called adventitious shoots for the same reason. A shoot produced in the axil of a leaf is an axillary bud, but since this is normal it is not called an adventitious bud.

Although a number of other species, such as dahlia, produce tubers, adventitious buds do not form on them and they are not looked upon as a means of natural vegetative reproduction. However, several tubers form on each plant and if nurserymen divide them so that each has a shoot bud from the crown, several plants can be produced from each parent.

Finally we must mention the process of root budding. This refers to the process of the differentiation of shoot buds on roots. This occurs in a number of trees, such as the Aspen. Shoot buds are produced on shallowly-growing roots and are, therefore, not far from the soil surface. Growth of the bud produces an upright growing shoot which pushes out through the soil usually several metres from the parent stem. Adventitious roots form at the base of the new shoot and the new plant becomes independent of the parent, although contact is still maintained through the original root. As you can imagine, this process, over a period of years, can lead to the establishment of a growing forest of Aspen.

SAQ 7.1

Which of the items in list A match with the statements in B?

List A

Potato.

Strawberry runner.

Bulbs.

Corms.

Rhizomes.

Root budding.

Tuberous root.

List B

1) Axillary buds function in the propagating system.

2) Consists of, or formed from, a horizontal stem.

3) Consists of a swollen storage organ.

4) Consists of, or formed from, a slender propagating system.

5) Requires formation of adventitious shoots for propagation.

7.3 Sexual reproduction

Flowers are the organs of sexual reproduction in the class Angiospermae, the flowering plants, whereas cones serve this purpose in the class Gymnospermae. Whilst in both classes, male and female gametes fuse to produce a zygote which develops into a seed, flowering plants show much more variation in the structure of their reproductive organs. We will be concerned mainly with flowering plants.

The basic structure of a flower is shown in Figure 7.2.

sepals and petals arranged in whorls

The various parts of the flower, such as sepals and petals, are all considered to be modified leaves and they are arranged in rings which are given the name whorls. The flower itself is carried on its stalk, the pedicel, and all the different parts of the flower are attached to the stalk via the pedicel. There is incredible variation in the shape and size of flowers, including differences in the number of the various parts and, very often, their fusion either with each other or with some other part of the flower. Flower structure is the major factor used in the classification of flowering plants. For this reason a whole catalogue of botanical terms has been generated so that each form of floral morphology can be accurately described. This is an absolute necessity for plant taxonomy, but not for plant physiology. We will use only those terms which are necessary to understand our story.

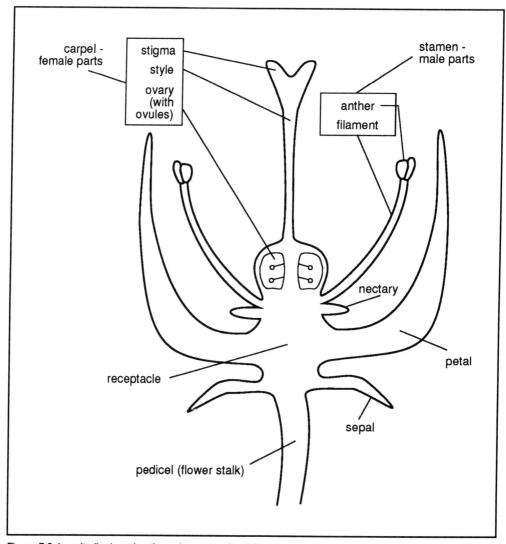

Figure 7.2 Longitudinal section through a generalised flower.

7.3.1. Formation of male and female gametes

stamens and
carpels

For reproductive purposes, the functional parts of the flower are the male stamens and female carpels. The petals and sometimes the sepals are coloured to attract insects but are not essential features and are absent in wind-pollinated plants.

Each stamen consists of a filament and an anther, the latter being the site of production of the pollen (Figure 7.3).

Each carpel consists of a stigma, style and ovary. The ovary is hollow and contains one or more ovules in which is formed an embryo sac (Figure 7.4).

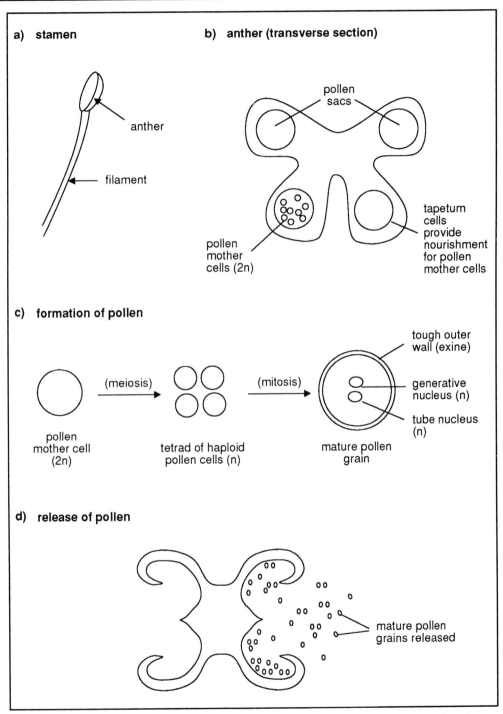

Figure 7.3 Production and release of pollen by the male parts of the flower. a) the stamen consists of a filament and anther; b) a cross section of the anther reveals four lobes each of which contains pollen mother cells; c) the pollen mother cells undergo meiosis to produce haploid cells which each divide once more by mitosis to produce mature pollen grains; d) at this stage the anther lobes break open releasing the pollen. (Note that each figure is not drawn to scale).

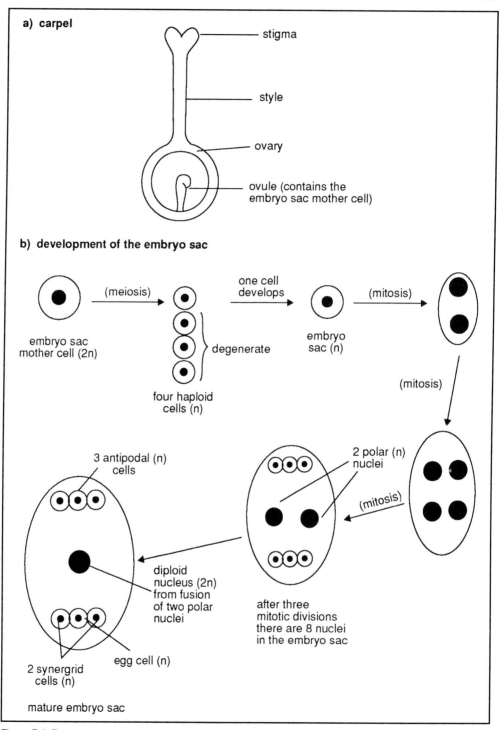

a) carpel

stigma

style

ovary

ovule (contains the embryo sac mother cell)

b) development of the embryo sac

(meiosis)

one cell develops

(mitosis)

embryo sac mother cell (2n)

four haploid cells (n)

degenerate

embryo sac (n)

(mitosis)

2 polar (n) nuclei

3 antipodal (n) cells

diploid nucleus (2n) from fusion of two polar nuclei

(mitosis)

after three mitotic divisions there are 8 nuclei in the embryo sac

2 synergrid cells (n)

egg cell (n)

mature embryo sac

Figure 7.4 Formation of the embryo sac in the ovule. a) The carpel consits of the style and stigma attached to the ovary. The ovary is hollow and one or more ovules develop within it, each ovule containing the embryo sac mother cell; b) the mother cell undergoes meiosis but three of the four cells die. The remaining cell divides by mitosis to produce eight nuclei which, after two of them fuse, become arranged as the three antipodal cells, the two synergids, an egg cell and the central dipoid nucleus. Note: n = haploid nucleus, 2n = diploid nucleus.

When released, each pollen grain contains two nuclei, the pollen tube nucleus and the generative nucleus. Pollination is completed once pollen alights onto the sticky surface of the stigma. This is followed by germination of the pollen with the formation of the pollen tube, which grows down the style and enters the ovary (see Figure 7.5). The tube nucleus follows closely behind the tip of the pollen tube. During this time the generative nucleus divides by mitosis to produce two haploid nuclei called sperm nuclei.

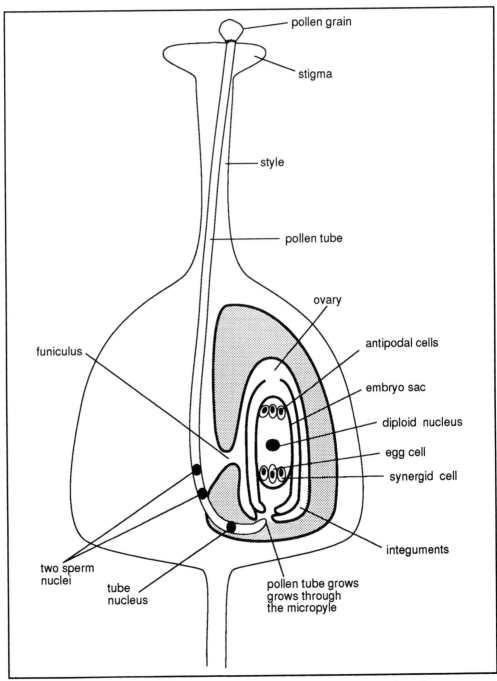

Figure 7.5 Growth of the pollen tube down the style to enter the ovule through the micropyle.

During the formation of the embryo sac the ovule tissue develops two layers which grow over the surface of the inner tissue to meet at the synergid/egg cell end of the embryo sac. These layers will eventually become the seed coat. However, they do not fuse when they meet but leave a small pore called the micropyle (Figure 7.5) through which the pollen tube grows.

micropyle

7.3.2 Fertilisation is a double event

Once inside the embryo sac, the pollen tube stops growing and its tip degenerates. The two sperm nuclei migrate out of the tube to be adjacent to the synergids and the egg cell. One of the sperm nuclei fuses with the egg cell, forming the diploid zygote. The other one migrates to the centre of the embryo sac and fuses with the diploid nucleus forming the triploid endosperm nucleus. The triploid nucleus undergoes many mitotic divisions to produce a large number of nuclei, all within the central portion of the embryo sac. Eventually cell wall formation occurs around the cells producing the multicellular food storage tissue, the endosperm.

7.3.3 Development of the embryo

The zygote undergoes a characteristic series of mitotic divisions which are shown in simplified form in Figure 7.6.

octet formation

globular
embryo

The early mitotic divisions produce a polar embryo, characterised by the large suspensor cell which actively absorbs food from the developing endosperm. The octet of cells is pushed deeply into the endosperm by the formation of the line of cells. The next division by the octet separates an outer layer from an inner one. This is an important stage because it separates the cells which will become the dermal tissue from the central ones which will give rise to the ground and vascular tissues. Subsequent cell divisions occur evenly and produce a spherical entity, the globular embryo, but thereafter the rates of cell division vary resulting in two protuberances which will eventually become the two cotyledons. Note that vascular and ground tissue are segregated at this stage. Subsequent development produces the torpedo-shaped embryo characterised by the delineation of the root - shoot axis.

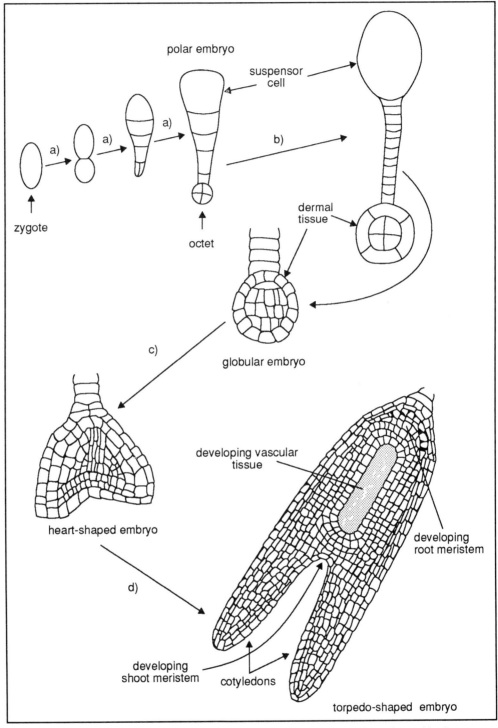

Figure 7.6 Major stages in the formation of the embryo. a) Formation of the polar embryo with a large suspensor cell at one end and an octet of cells at the other, b) division of octet to produce 16 cells in two layers, with subsequent development producing the globular embryo, c) uneven rates of cell division convert the globular-shaped embryo into a heart-shaped embryo and development continues to eventually produced the torpedo-shaped embryo shown in d). (Not drawn to scale).

7.3.4 Formation of the seed

During the growth of the embryo, the two integuments of the ovule fuse to become the seed coat, the testa. The endosperm has continued to acquire food and pass it to the embryo but we can recognise two types here. In some species, such as castor bean and the lime tree, the cotyledons are thin and papery and the endosperm persists as the food store in the mature seed (Figure 7.7).

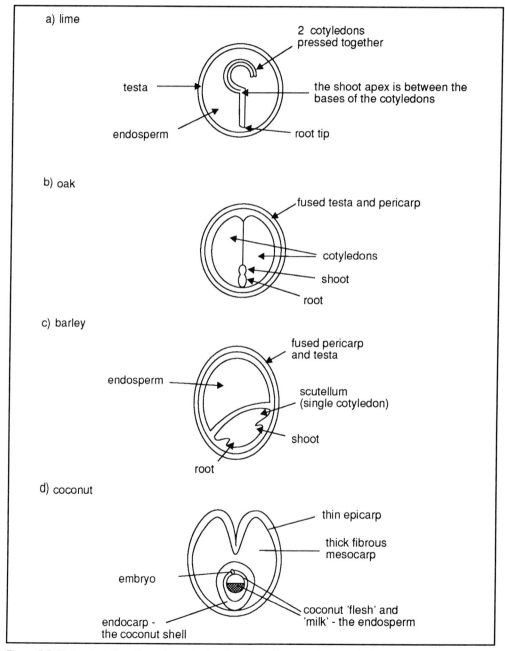

Figure 7.7 Stylised sections through seeds showing a) persistent endosperm in the seed of the lime tree, b) development of the cotyledons as the food store in oak, c) persistent endosperm of a monocotyledon seed, eg barley and d) production of liquid and solid endosperm in coconut.

In species such as oak, pea and bean, the endosperm passes all of its food to the cotyledons so that, at maturity, the endosperm is virtually non-existent but the cotyledons are large and fat, and account for most of the seed's volume.

7.3.5 Formation of the fruit

We have seen that the ovule gives rise to the seed and that the surface layers of the ovule becomes the seed coat. What about the fruit? By definition, the fruit develops from the ovary wall (the pericarp) and because the ovules are inside the ovary, the seeds are borne within the fruit. As you know, the fruit can take a number of forms. In many cases the pericarp becomes fleshy as in the grape and this is known as a berry. In others, such as radish and sunflower, it is thin and dry and the fruit is called an achene. In contrast, the pericarp may be very hard, as in a nut or it can be both fleshy and hard. This striking phenomenon is called a drupe. In drupes the pericarp differentiates into three zones. The outer layer (exocarp) becomes a thin skin, the middle layer (mesocarp), becomes fleshy and the innermost layer (endocarp), becomes as hard as a nut.

berries
achenes

drupes

∏ The legend to Figure 7.7 suggests that the diagrams show seeds of lime, oak, barley and coconut. In the light of the above definition of a fruit, is this legend correct?

It is correct only in the case of the lime seed. Diagrams b), c) and d) are fruits.

∏ Can you think of an example of a drupe?

They are, actually, quite common, and include the cherry, peach, plum and olive.

SAQ 7.2

The diagram below shows a stylised fruit with the labelled parts assigned a letter. Which of the statements below apply to each of the labelled parts of the diagram?

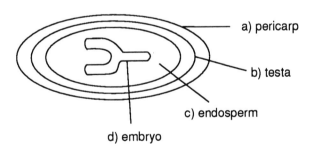

a) pericarp

b) testa

c) endosperm

d) embryo

1) Contains the diploid chromosome number.

2) Contains the haploid chromosome number.

3) Contains the triploid chromosome number.

4) Expresses maternal characteristics only.

5) Expresses paternal characteristics only.

6) Expresses both maternal and paternal characteristics.

7) Is derived from the ovule only.

8) Is derived in part from a product of pollen.

| SAQ 7.3 | Bearing in mind that an ovary can contain many ovules, draw a diagram similar to that of Figure 7.4 a) to show the arrangement of ovules that you might expect in a pea carpel. |

It should be obvious from the above that this has been a very simplified account of the morphology of flowers, seeds and fruits. It is sufficient for our purposes, but if you require more detail on this (fascinating) topic it can be found in a number of botanical texts, a particularly good one being Animal and Plant Biology, Vol 1, A.E. Vines and N. Rees, published by Pitman.

7.4 Seed development, dormancy and germination

extensive biosynthesis during seed development

water loss and dormancy

The development of a seed is characterised by intense metabolic activity. We saw in Chapter 5 that seeds constitute the most active sinks in the plant and that sucrose and amino acids are very actively transported into them. These raw materials are partly converted into the structure of the developing seed and partly into the food store. Thus this is a time of very high anabolic activity and the biosynthetic pathways leading to starch, proteins and fats are very prominent. Towards the end of seed development the funiculus, which attaches the ovule to the ovary, disintegrates thus severing the link between the parent and offspring. Whereas many biosynthetic reactions continue from this time, there is no further increase in seed dry matter. At this stage, the seed begins to lose water and dry out until it has a moisture content of 10-15%. At this level of moisture, seed metabolic activity is very low. The seed of most species is now in a state called dormancy, which refers to an inability to grow when provided with suitable environmental conditions such as water, O_2 and a suitable temperature. This imposition of dormancy is very important because it prevents precocious germination. Bearing in mind that the function of a seed is the dispersal of a propagating unit, it would be a disadvantage to a species if its seeds germinated on the parent plant. In many cases ABA accumulation during the later stages of seed development appears to be instrumental in preventing precocious germination, but the ABA does not seem to persist and is usually present in negligible quantities when the seed is shed. Why, then do seeds not germinate when placed in a suitable environment?

7.4.1 There are many reasons why seeds are dormant

seed coat may impose dormancy

The seeds of many families, such as the Leguminosae and Geraniaceae, possess testae which are either impermeable to water, prevent oxygen uptake or simply offer mechanical resistance to penetration by the root. If the seed coat is removed and the embryo cultured in moist warm conditions, germination occurs showing that the embryo *per se* was not dormant. However, the presence of the seed coat in intact seeds imposes a considerable delay in germination. Hard seed coats are slowly degraded by weathering and by soil micro-organisms and their effect is gradually overcome. Assuming seeds are shed in autumn, by the time the seed coat is degraded winter has arrived when soil temperature will be too low for germination. Note that plant growth virtually stops at 6°C and, in northern Europe, soil temperatures are usually below this value from mid-November to mid-April. Thus, although seed testae may have been broken down by December or January, germination will not occur until late April because of the soil temperature. Knowing the effect of the testa, nurserymen can obtain germination when desired by physically abrading the seed, the process being called scarification.

scarification

inhibitors of germination

It has been known for decades that many species produce chemical inhibitors of seed germination and an incredibly wide range of compounds has been identified as having these properties. These include coumarin, parasorbic acid, ferulic acid and ABA. Examination of the isolated embryos from species which belong to this group, however, reveals in most cases that the embryo itself is not dormant. Evidence suggests that the seed coat may reduce the rate of leaching of these compounds and that by the time this has occurred the temperature conditions are no longer conducive to growth.

Having stressed the retarding effect of low temperature on germination, it is perhaps strange to find that, in many species, low temperature exposure is required for germination, eg in apple, pine, lime and grape. In most of these species the isolated embryos are dormant so this is not generally regarded as a seed coat phenomenon.

SAQ 7.4

In an experiment to study the effect of cold treatment on apple, seed were exposed to a particular temperature range for 85 days and then transferred to 22°C for 21 days, at which time germination was scored. This experiment was done in two ways. When the seeds were moist during the 85 day exposure the results below were obtained:

Temperature of 85 day exposure (°C)	Subsequent % germination at 22 °C
0	0
1	25
3	80
5	90
7	60
9	5

When the seeds were kept dry during the 85 days exposure, no germination occurred on transfer to 22°C.

What is your interpretation of these observations?

It is considered that at 5°C a metabolic reaction (or reactions) is occurring preferentially and that this removes the block to germination. In certain cases, the evidence suggests that gibberellin synthesis may be reactivated by chilling and that an inability to synthesise this hormone constitutes the germination block. In species which require chilling for seed germination it is considered that the requirement is usually met by, for example, early January in England. Of course, soil temperatures would not allow germination to begin until much later in the year.

SAQ 7.5

The above discussion of seed dormancy shows that a wide array of mechanisms have been evolved by plants. How can we explain this? Why is there not a single universal mechanism? There is a hint in the response, should you need it.

7.4.2 Some seeds require light before they will germinate

To complete the discussion of seed dormancy we should mention the light-requiring seeds, examples of which include tobacco (*Nicotiana spp*), foxglove (*Digitalis purpurea*), dock (*Rumex crispus*) and lettuce (*Lactuca sativa*). Initially, it is a little difficult to account for a requirement for light for seed germination but recent work examining the penetration of light into soils allows a rationalisation. Experiments show that light can

light penetration in soils

penetrate to a considerable depth in soils, which may be up to 10cm in sandy soils. Obviously, penetration is less in clay soils but the results do indicate that soil should not be considered as an absolute light-barrier.

The explanation suggested for the phenomenon of light-requiring seeds is that it is a mechanism which prevents the germination of deeply-buried seed. Seeds obviously have a finite amount of stored food and if a seed finds itself buried deeply its reserves may run out before the shoot penetrates the surface. A mechanism which prevents germination under these conditions will be selected for. Such seed will lie dormant until the soil is disturbed and they are moved closer to the surface.

7.4.3 A reservoir of weed seeds is present in the soil

Whereas we are concerned about seed production in crop species we must not forget about weed seeds. As with all plants, the seeds of weed species are important for the process of the colonisation of new habitats and also to protect the species against unfavourable environmental conditions via dormancy. However, because of their greater seed production (Table 7.1) weeds may have a considerable competitive advantage over crop species.

Weed	Common Name	Seeds per plant (approximate)
Triticum aestivum	wheat	90
Capsella bursa-pastoris	shepherd's purse	3500
Papaver rhoeas	poppy	14000
Stellaria media	chickweed	15000

Table 7.1 Seed production by wheat and some major weeds.

This advantage occurs not only because of a greater annual production but also because of the persistence of weed seeds in the soil. The seed population in soil is reduced by germination, by decay and by being eaten by herbivores. Despite this, it is estimated that there are up to 75000 viable seeds m^{-2} of arable soil. When this is compared with the normal sowing density for wheat of approximately 400 m^{-2} the extent of the advantage becomes evident. Thus, seed biologists are interested in maximising seed performance in crop species while minimising it in weed seeds.

7.4.4 Establishment is the goal of germination

water and O_2 absorption

A seedling is said to be established when it has begun to photosynthesise and is no longer dependent on the reserves laid down in the seed. When a non-dormant seed is exposed to water, O_2 and a temperature above 6°C, water absorption occurs and the embryonic tissue regains a moisture content of 80-90%. O_2 absorption occurs as the seed swells, this being required for the increased respiration which accompanies germination. Most species show root development before shoot development. The root elongates, emerges from the seed and grows down into the soil, producing root hairs and branch roots, described in earlier chapters. Once the root has made effective contact with the soil, the shoot commences growth and shows either epigeal or hypogeal development. We need to introduce some terms to explain the differences between these two. Examine Figure 7.8 before reading on.

epigeal or hypogeal development

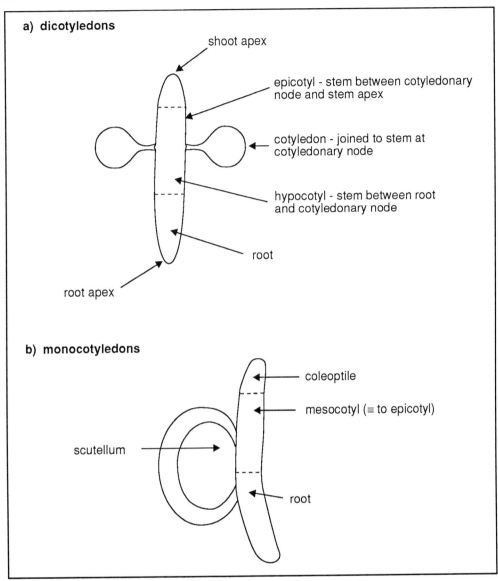

Figure 7.8 Diagram of young seedling to show the position of a) the hypocotyl and the epicotyl in dicotyledons, and b) the mesocotyl in monocotyledons.

Epigeal and hypogeal refer to the position of the cotyledons relative to ground level when germination is complete. In epigeal germination the cotyledons are brought above the ground but in hypogeal germination they stay below the soil surface.

∏ Examine Figure 7.9. Which of them show epigeal and which hypogeal germination?

Epigeal germination is shown by a) and b), and hypogeal by c) and d).

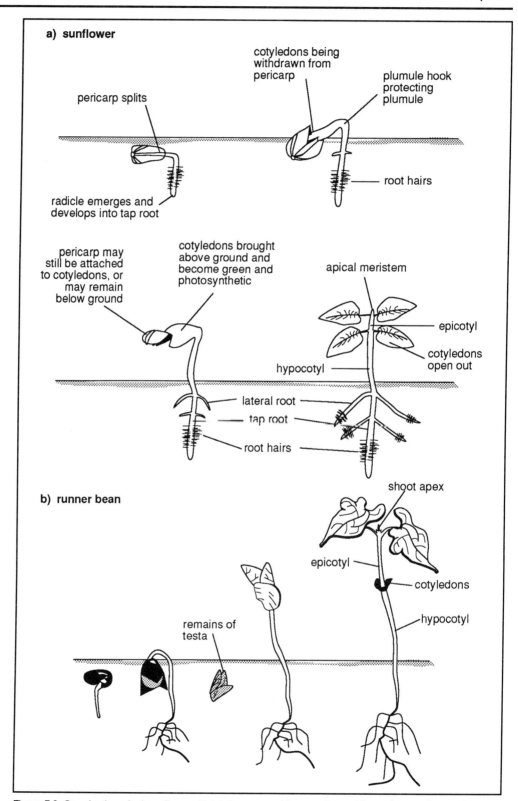

Figure 7.9 Germination of: a) sunflower, *Helainthus annus*, b) runner bean, *Phaseolus coccineus*. (Cont'd).

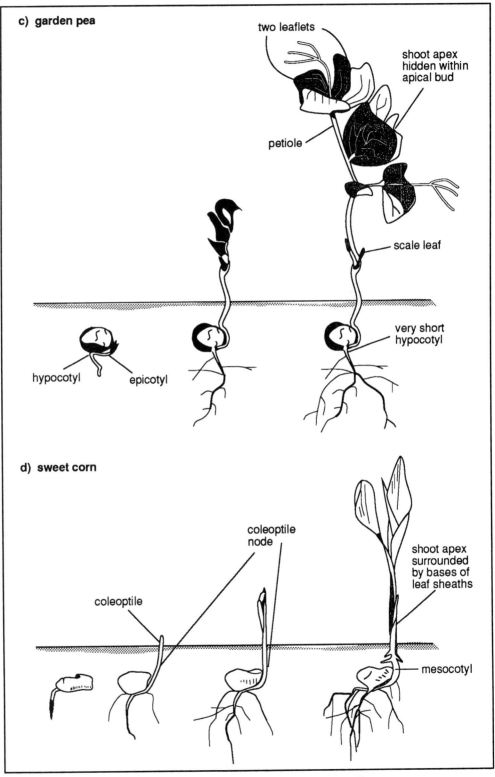

Figure 7.9 Germination of: c) garden pea, *Pisum sativum* and d) sweet corn, *Zea mays*.

∏ Can you observe any difference between germination in a) and b)?

Whereas both show epigeal germination, in a) the cotyledons become functional leaves, whereas in b) they do not.

Notice that, if the epicotyl extends during germination, germination is hypogeal, whereas if the hypocotyl extends, germination is epigeal. Although the function of the hypocotyl and root are different, it is usually difficult to decide where the dividing line is between them, without looking at cross sections.

Biochemically speaking, germination is the opposite to seed development because in germination the food reserves are broken down and mobilised to the growing root and shoot. Obviously, root and shoot growth must be coordinated with the breakdown of reserves and there is evidence in cereals and other species that gibberellins are involved in this coordination. In cereals, starch and protein reserves in the endosperm are broken down by the action of amylases and proteases secreted from the aleurone layer, which surrounds the endosperm. In species such as barley and wheat the embryo can be easily removed from the seed by excision (Figure 7.10) and in such embryo-less seeds the digestive enzymes are not produced.

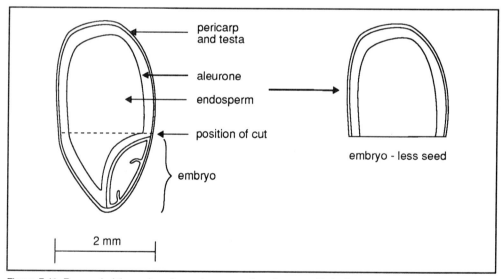

Figure 7.10 Removal of the embryo to produce an embryo-less seed in barley.

The process of synthesis and secretion of the digestive enzymes in embryo-less seeds is triggered by gibberellin application but by no other hormone. It was subsequently shown that GAs are produced by the embryo of the germinating seedling and that this production precedes the production of α-amylase (Figure 7.11).

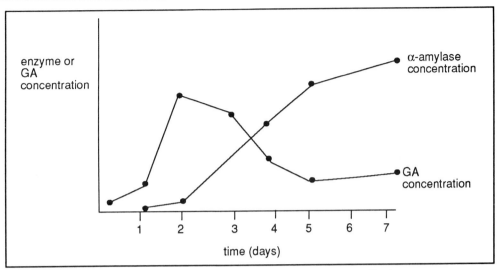

Figure 7.11 Changes in GA and α-amylase concentrations during germination of barley.

Using actinomycin D, an inhibitor of RNA synthesis, and cycloheximide, an inhibitor of protein synthesis, it was shown that, in germinating barley, GA application causes the synthesis of messenger RNA molecules which code for the digestive enzymes. This is, therefore, an example of a GA action which involves activity at the gene level; GA stimulates expression of genes coding for digestive enzymes.

foodstores are mobilised Thus, as growth begins at the start of germination, gibberellins are synthesised by the embryo and transported to the aleurone layer. The digestive enzymes are synthesised and released into the endosperm where they act to degrade starch and protein. Sucrose and amino acids are then transported to the growing embryo under the influence of the increasing sink activity generated there (Figure 7.12).

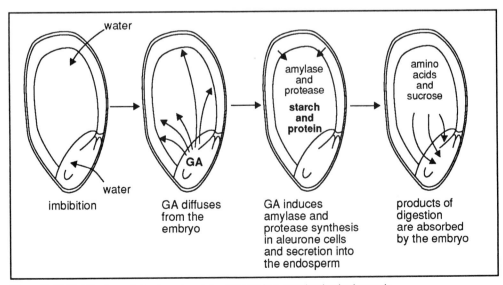

Figure 7.12 Mobilisation of starch and protein reserves in a germinating barley grain.

seed vigour   The percentage of planted seeds which become established is important to a grower and the actual rate at which germination occurs is a major factor in this. The term seed vigour is used to encompass all the factors which contribute to quick establishment and the increase in vigour is one of the main targets for research in this area.

We are now ready to look at how flowering is controlled.

7.5 The control of flowering

importance of the duration of the photoperiod

The control of flowering had puzzled plant physiologists for many years before two Americans, Garner and Allard, provided the first clue in the early 1920s. They and others had noted that individual species tended to flower at the same time each year irrespective of the amount of vegetative growth they exhibited. This was particularly striking with plants such as soybeans (*Glycine max*) which had been germinated early in the year in a greenhouse and transplanted out of doors when the danger of frosts was past. Although they grew much taller than plants sown later out of doors, they all flowered at the same time. Garner and Allard showed that the controlling factor was photoperiod, the length of the daily light period, and that plants could be divided into three groups:

- short-day plants (SDPs) - flower when exposed to less than a critical number of hours of light;

- long-day plants (LDPs) - flower when exposed to more than a critical number of hours of light;

- day-neutral plants (DNPs) - flowering is independent of photoperiod.

Subsequent work has shown that some species remain permanently vegetative if kept under unfavourable daylength conditions and are called obligate photoperiodic plants. Other species show earlier flowering under long days or short days but will ultimately flower under unfavourable conditions and are called quantitative photoperiodic plants. Examples of these groups are shown in Table 7.2.

Short-day plants	Long-day plants
Obigate	Obligate
Nicotiana tabacum (tobacco)	*Avena sativa* (oats)
Fragaria spp (strawberry)	*Lolium temulentum* (rye-grass)
Coffea arabica (coffee)	*Trifolium spp* (clover)
Quantitative	Quantitative
Cannabis sativus (hemp)	*Antirrhinum majus* (snapdragon)
Gossypium hirsutum (cotton)	*Beta vulgaris* (sugar beet)
Oryza sativa (rice)	*Lactuca sativa* (lettuce)

Table 7.2 Examples of short-day and long-day plants.

SDPs include those which are indigenous to regions of low latitude, that is close to the equator. LDPs plants are native to temperate regions which experience long days in summer.

∏ SDPs also occur in temperate regions such as Northern Europe. When do you think SDPs flower in the UK?

Theoretically, SDPs could flower in spring or autumn when the days are short. However, plants grown from seed outside, will be very small during spring and will, therefore, usually flower in autumn. We will return to spring-flowering plants later.

SAQ 7.6

1) Examine Figure 7.13 and state the critical daylength for flowering in *Pharbitis nil* and *Sinapis alba*.

2) *Xanthium* is a SDP and has a critical day length of 15.5 hours, whereas *Hyoscyamus niger* is a LDP with a critical daylength of 11 hours. State the response of each plant when kept under controlled light/dark periods of a) 16 hours/8 hours, b) 12 hours/ 12 hours and c) 8 hours/16 hours.

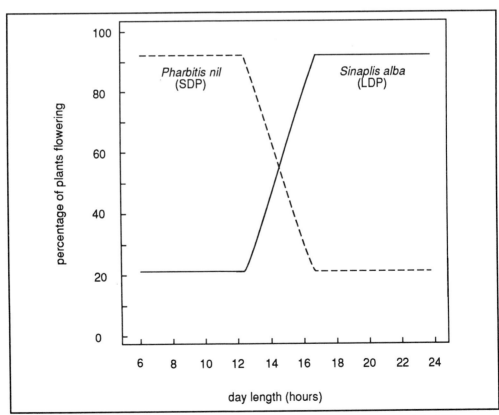

Figure 7.13 Effect of day length on flowering in the LDP *Sinapis alba* (white mustard) and the SDP *Pharbitis nil* (Japanese morning glory). Redrawn from Wilkins M.B., Plant Watching, Roxbury Press, 1988.

7.5.1 Plants measure the length of the night period

plants are
governed by
the duration of
the dark period

The foregoing section may have given the impression that plants detect the length of the light period but experiments show that it is actually the length of the night that is detected. The experiments designed to show this could not be carried out using natural daylight because of the limitation of the 24h day. Lengths of night and day can only be changed independently of each other in a controlled environment facility. Garner and Allard did not have access to such a facility and the experiment was first performed by Hamner in 1940. Figure 7.14 shows the results of his experiments.

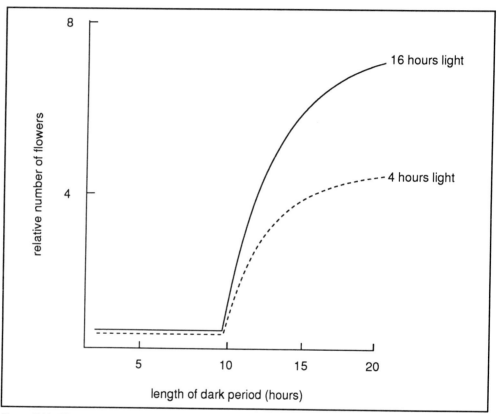

Figure 7.14 Effect of various lengths of dark period, in association with constant 16 hours or 4 hours of light, on flowering in soybean (*Glycine max*).

With both photoperiods flowering was induced only on exposure to dark periods of greater than 10 hours.

Hamner did the converse experiments in which he maintained plants on a fixed 16 hour dark period and increased the length of the light period. These showed that flowering increased at exposures up to 12-14 hours and then declined. Thus for flowering in soybean, the requirements are for a daily dark period of more than 10 hours and a daily light period which must not exceed 12-14 hours. Of course, under natural conditions dark periods of 10 hours or more can only be accompanied by light periods of 14 hours or less. Thus, for flowering in soybean, the dark period is the critical one and this has been found to be the case in other SDPs and LDPs. For this reason it could be argued that LDPs and SDPs should be renamed short night and long night plants respectively. However, this renaming, although quite correct, has never caught on.

Having reached the conclusion above that plants can detect the length of the dark period let us now ask how this might be achieved.

7.5.2 Study of germinating lettuce leads to the discovery of phytochrome

In order to continue our story we must digress to re-examine something we mentioned earlier, namely the effect of light on the germination of lettuce, especially the variety Grand Rapids. We saw in Chapter 2 that any process involving light can be studied by determining the action spectrum of the process. When this was done with these lettuce seeds it was found that in the visible part of the spectrum only red light (600-700 nm) was effective in enhancing germination. An exposure to far red light (700-800 nm) which acts in the same way as dark when used alone, was unexpectedly shown to counteract the effect of previous red exposure. Table 7.3 shows the result of several such alternations of light exposure.

	Treatment	% Germination
1)	dark	7
2)	red	70
3)	red/far-red	6
4)	red/far-red/red	74
5)	red/far-red/red/far-red	6
6)	red/far-red/red/far-red/red	73

Table 7.3 Effect of alternate red and far-red light exposure on the germination of Grand Rapids lettuce. The duration of exposure to light was 5 min using a 40 watt incandescent light bulb after which seeds were kept in darkness for 3 days and % germination measured.

It was apparent that lettuce germination could, in effect, be switched on and off by exposure to red or far-red light and a new pigment was proposed to account for this. The name phytochrome was coined for this pigment which was predicted to exist in two interconvertible forms called P_r and P_{fr} (Figure 7.15).

phytochome

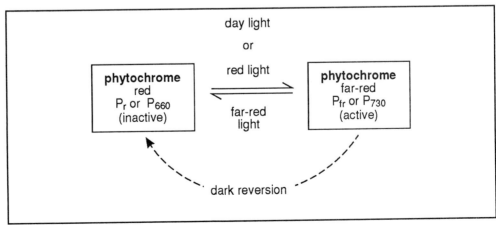

Figure 7.15 Effect of red and far-red light on interconversion of the two forms of phytochrome.

The pigment has subsequently been isolated and accurate absorption spectra obtained (Figure 7.16). It is protein in nature and consists of a chromophore attached to an apoprotein portion.

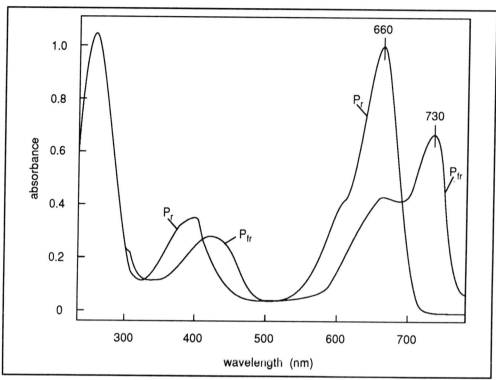

Figure 7.16 Absorption spectra of the two forms of phytochrome. P_r is the spectrum obtained after exposure to 730 nm light; P_{fr} is the spectrum obtained after exposure to 660 nm light.

To summarise, phytochrome occurs in two forms, now generally referred to as P_{660} and P_{730}. Red light acts in the same way as day (white) light and far-red acts in the same way as dark although dark reversion is a slow process. Red light converts P_{660} into P_{730}, far red light converts P_{730} into P_{660}.

The structure of the phytochrome chromophore is shown in Figure 7.17.

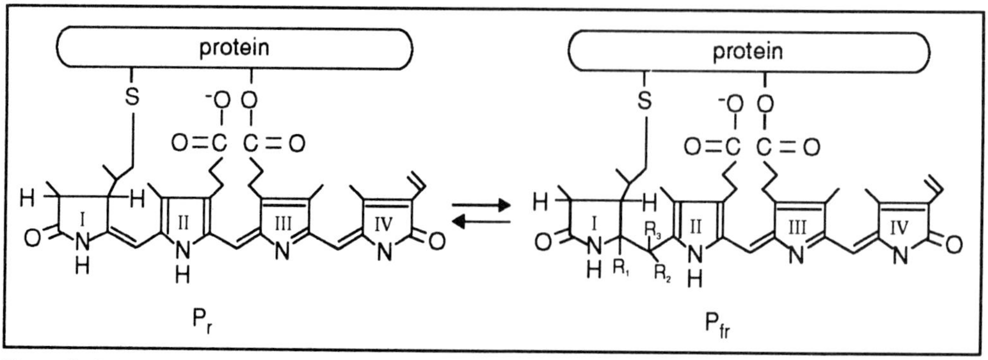

Figure 7.17 The phytochrome chromophore is a linear tetra pyrrole linked to the apoprotein via the I and III ring. Note that P_r and P_{fr} differ in substituents between rings I and II.

The interpretation of Table 7.3 is that P_{730} must be present for germination to occur. Exposure to far-red light removed P_{730} which prevented germination. In all phenomena involving phytochrome, the critical factor is the amount of P_{730} present, P_{660} is without action.

7.5.3 Phytochrome controls many developmental processes

Phytochrome is now known to cause over one hundred responses in plants in addition to its effect on germination. Some of these are listed in Table 7.4.

i)		Stimulation of opening of plumular hook
ii)		Inhibition of stem extension growth
iii)		Stimulation of leaf growth and differentiation
iv)		Stimulation of xylem differentiation
v)		Stimulation of chloroplast development
vi)		Stimulation of chlorophyll, carotenoid and anthocyanin biosynthesis
vii)		Stimulation of bud dormancy
viii)		Stimulation of formation of epidermal hairs

Table 7.4   Plant responses to phytochrome (P_{730}).

7.5.4 Phytochrome is also involved in the control of flowering

night interruption experiments

The involvement of phytochrome in the control of flowering was discovered as a result of what are called night interruption experiments. We saw in Figure 7.14 that soybean is a SDP and has a critical dark period of 10 hours. Thus, if exposed to a regime of 12 hours dark/12 hours light we would expect it to flower, and that was shown to be the case. However, if the 12 hours of dark was interrupted after 6 hours by one minute of light the plants failed to flower.

This provided a very useful opportunity for further investigation because it becomes possible to determine the action spectrum of light needed for this phenomenon. This action spectrum was shown to have the same shape as that of P_r in Figure 7.16. You can imagine what experiments were done next. It was very simple to show that 1 minute of red light had the same effect as 1 minute of white light and also that 1 minute of far-red light applied after the 1 minute red counteracted its effect. The action spectrum of the second light treatment showed the same curve as that of P_{fr} in Figure 7.16. This suggested that phytochrome was involved. Similar types of experiment have been done with LDPs, where long nights prevent flowering but night breaks with white or red light counteract this.

Let us make sure you understand these points with an SAQ.

SAQ 7.7

Soybean is a SDP and its critical day length is 14 hours.

Spinach is a LDP and its critical day length is 13 hours. The grid below shows a number of light:dark cycles with and without short breaks of light and dark. Fill in the boxes with F for flowering or V for vegetative for the two species.

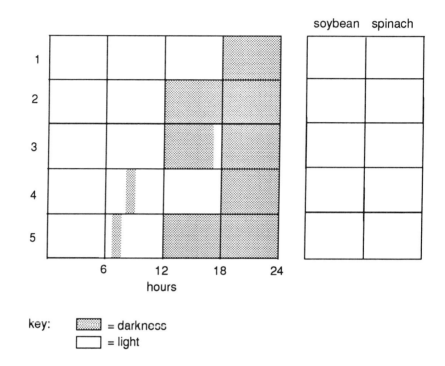

Night interruption phenomena with white light, because of its action, causes the formation of P_{730}, and suggests that P_{730} is inhibitory to flowering in SDPs in the dark, but stimulatory to LDPs in the dark. This has led to the proposal of the 'hour-glass' model for the way in which plants measure the length of the dark period. At the end of the day phytochrome is in the P_{730} form and the hour glass is set. During darkness P_{730} gradually reverts to P_{660} (the sand falls into the bottom compartment). If the dark period is short P_{730} persists throughout and SDPs will be vegetative, whereas LDPs will flower. If the dark period is long enough for all the P_{730} to revert to P_{660} then SDPs will flower but LDPs will not. We have implied in the above that P_{730} levels must fall to absolute zero but the model could just as easily incorporate a threshold level of P_{730}, as shown in Figure 7.18.

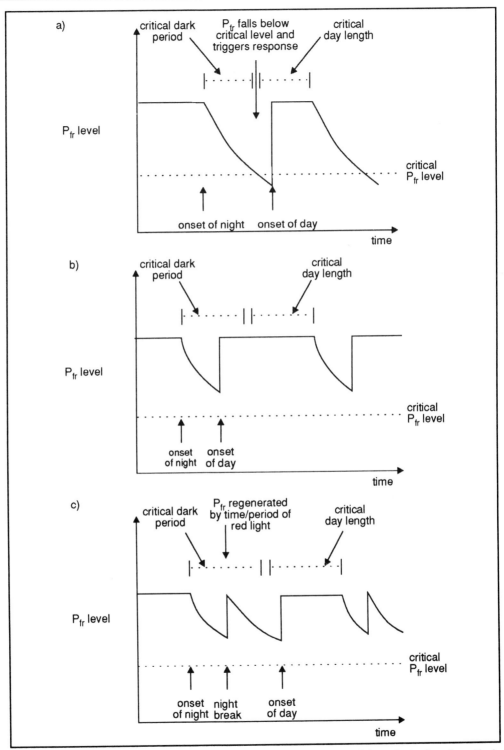

Figure 7.18 Changes in the level of P730 in three photoperiodic treatments: a) days shorter than the critical day length (long nights), b) days longer than the critical length (short night), c) as in a) but with the night interrupted by a brief period of red light. (After Irene Ridge Ed. Plant Physiology, Hodder and Stoughton, 1991).

7.5.5 The hypothesis tested

The measurement of phytochrome is quite difficult because of the masking effect of chlorophyll, which absorbs in the same part of the spectrum. Early work was limited to tissues with very low chlorophyll levels, such as plants grown in the dark or in low light, but methods are now available for measuring the levels of P_{660} and P_{730} in green tissue. The description of these methods is beyond the scope of the present text and readers are referred to the text by H. Smith; Phytochrome and Photomorphogenesis, McGraw Hill (1988).

measurement of P_{660}/P_{730} not consistent with the hypothesis

When the levels of P_{730} were measured scientists found that reversion to P_{660} occurred more quickly than expected and that it is virtually complete within 5 hours. This makes it difficult to see why critical dark periods are as long as they are. Why does soybean need 10 hours of dark if P_{730} disappears after 5 hours? There is no agreement at the moment as to the answer to this question, but two important considerations are a) rhythms of sensitivity to light and b) the production of a flowering hormone. Let us examine these two ideas briefly.

7.5.6 Endogenous rhythms in flowering

Numerous scientists, Bünning in particular, have drawn attention to the existence of persistent rhythms in plants. A particularly clear example of this is the diurnal movement of leaves of the runner bean (*Phaseolus multiflorus*), in which the primary leaves rise during the early part of the day, fall during the later part and reach a minimum position during the night. When this movement is viewed with time-lapse photography it looks as if the plant is flapping its leaves. These movements are controlled by turgor pressure changes in the cells at the base of the leaf. Bünning proposes that plants show a rhythm of photoperiodic sensitivity during which they alternate between a photophile phase, when light is favourable to flowering and a photophobe phase, when light is inhibitory. Bünning's system would suggest that in soybean, for example, the photophobe phase lasts for 10 hours and the photophile stage for 14 hours. Thus soybean will flower when the natural daylength fits into this daily rhythm. A very detailed experiment of Hamner's provided some evidence in favour of Bünning's hypothesis. This is shown in Figure 7.19.

photophile and photophobe periods

flowering is geared to 24 hour cycle in soybean

The results shown in Figure 7.19 strongly suggest that flowering in soybeans is geared to a 24 hour cycle and that when the cycle length is 24, 48 or 72 hours the endogenous photophile and photophobe phases will correspond with the environmental light and dark periods, but will not do so with other periods. Promising though these results were, they were shown not to apply to a number of other plants such as *Xanthium* and *Kalanchoe*, so the question of the involvement of rhythms of sensitivity remains unresolved.

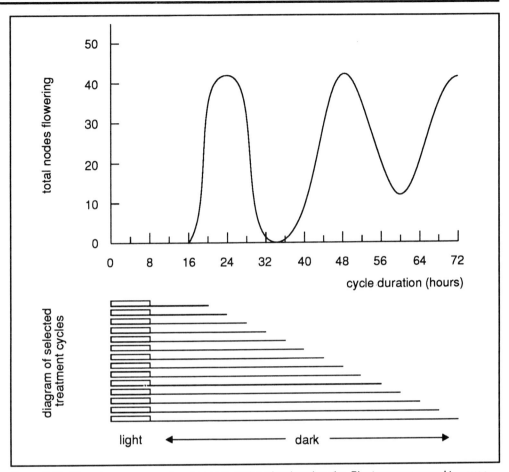

Figure 7.19 Flowering response of soybeans to cycles of various lengths. Plants were exposed to seven cycles of light and dark. Plants received 8 hours of light in all cases but various dark periods. (Modified from Environmental Control of Plant Growth, K.C. Hamner, Academic Press 1963).

7.5.7 Is there a flowering hormone?

photoperiodic stimulus perceived by leaves (florigen)

It has been demonstrated unequivocally that the leaves of plants perceive the photoperiodic stimulus, but it is the stem and branch apices which respond by flowering. This suggests the production of a flower-promoting stimulus in leaves and its transport to the apices. The Russian scientist, Chailakhyan, coined the term florigen for this 'flowering principle' but although it has been sought for over 60 years no one has been able to isolate and identify it.

Xanthium plants will flower once they have been exposed to one photo-inductive cycle, ie 24 hours in which the dark period satisfies the requirement for the species in question. All plants need several days before the initiation of flowers can be confirmed and this is usually done by dissection of shoot apical buds. *Xanthium* is particularly useful because it will flower when exposed to just one photo-inductive cycle. In these types of experiment, development between the end of the photo-inductive period and the confirmation of flowering must be under non-photo-inductive conditions.

| SAQ 7.8 | How could you demonstrate that the leaves but not the stem or apex perceive the photoinductive exposure? |

In the context of SAQ 7.8 it is quite astounding that with *Xanthium* only one-eighth of a leaf is required for photo-induction to be successful (Figure 7.20).

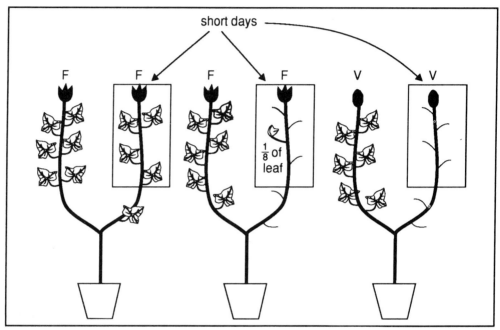

Figure 7.20 Effects of exposing leaves to light on flower development in *Xanthium*. Two-branched *Xanthium* plants allow two portions of a plant to be treated differently. The boxed branches represent the branch which is exposed to light. Flowering (F) occurs on both branches provided that the photo-induced plant retains one-eighth of a leaf. If all leaves are removed the branches remain vegetative (V) (After K.C. Hamner, Cold Spring Harbour Symposia *10*, 49 1942).

Further evidence comes from experiments in which branches or leaves from photo-induced plants are grafted to non-induced plants. These types of experiment are very important because they can be done using inter-specific grafts between different SDPs, between different LDPs and also using 'crosses' between SDPs and LDPs (Table 7.5).

Photo-induced	Non-induced	Response
donor	*receiver*	
SDP	SDP	flowering
LDP	LDP	flowering
SDP	LDP	flowering
LDP	SDP	flowering

Table 7.5 Grafts between species, even when one is a SDP and the other a LDP, result in flowering if the grafted (donor) leaf or branch has been photo-induced.

Why are the experiments described in Table 7.5 so important? What other experiment might you set up to test you conclusion?

model
involving
florigen not
proven

The various grafting experiments suggest that florigen is a universal flowering hormone and they allow us to return to our time-measurement problem. The hour glass model proposed that the time-measuring device was the time required for the removal of P_{730} but analysis showed that P_{730} disappeared more quickly than expected and left a period of time unaccounted for. It is possible, in SDPs, that florigen synthesis accounts for the extra time, if we propose that it can only be synthesised in the dark and in the absence of P_{730}. Thus, a finite period of time after the removal of P_{730} is required to synthesise an above-threshold amount of florigen. Unfortunately this simple hypothesis cannot be tested because, as mentioned above, no one has managed to isolate florigen. This raises numerous questions about its nature and indeed existence which are beyond the scope of this text and interested readers are directed to the text, 'Plant Physiology', by L Taiz and E Zeiger, Benjamin Cummins, 1991, for further discussion. For our purposes, the inability to measure florigen means that we cannot test hypotheses which involve it, which is very frustrating.

To complete our story, we should mention the time-measuring system in LDPs. It had originally been considered that LDPs required P_{730} to be present throughout the whole of the dark period and it would seem reasonable to suggest that this was required for florigen synthesis but we know that P_{730} is converted to P_{660} much more quickly than expected. Thus there is likely to be part of the critical short night when P_{730} is not present. At the moment there is no simple way to explain these facts.

Before we leave this problem we should examine the action of the known plant hormones in flowering. Cytokinins, abscisins and ethylene can all induce flowering under non-inductive photoperiods in, at least, one species each. Gibberellins, on the other hand, can do so in approximately 20 species, some of which are LDPs while others are SDPs. Whereas these points are interesting and, in some cases, commercially important (for example, the effect of ethylene on pineapples), the conclusion is that most species do not respond to the known hormones by flowering. It has been suggested that flowering might be controlled, not by a single substance, but by a complex interaction between several substances. In this hypothesis, the growth hormones would contribute to the complex but, as before, there is evidence both for and against this and so the position is unresolved.

7.5.8 Some species require to be chilled before they will flower

winter and
spring varieties
of wheat

It has been known for many years that certain cereals, such as barley and wheat, can be divided into two groups. Varieties called spring varieties are normally sown in spring to germinate and flower after the production of a fixed number of leaves. Other varieties, called winter varieties, if planted in spring, produce more than twice as many leaves before they flower and the plants do not usually produce mature fruit because of the onset of winter. If winter varieties are planted in autumn, they germinate and overwinter as small seedlings and flower and fruit in the second year, having produced the same number of leaves as the spring variety. The stimulus for the winter variety to do this is exposure to the cold of the winter. Winter chilling causes winter varieties to behave as spring varieties and this is indicated by the name of the phenomenon,

vernalisation

vernalisation. Numerous other species will flower *only* if they have been exposed to chilling and, although they do not always have spring varieties to simulate, the phenomenon is still called vernalisation. Table 7.6 lists some examples.

Beta vulgaris	sugar beet
Brassica oleracea	cabbage
Apium graveolens	celery
Digitalis purpurea	foxglove
Primula vulgaris	primrose
Cheiranthus cheirii	wallflower
Lolium perenne	perennial rye-grass

Table 7.6 Plants requiring a cold treatment before they will flower.

The list in Table 7.6 can be divided into two groups. Primrose, wallflower and rye-grass are perennial plants and flower each year having been chilled during the previous winter. The remainder are biennials, plants which grow vegetatively in their first year and then flower and fruit in their second year, after which the plant dies.

Many plants which require chilling are also photoperiodic, (eg perennial rye-grass). This species is a long-day plant, 12 hours being the critical time and it flowers during the middle of the year in Northern Europe.

SAQ 7.10

Many varieties of *Chrysanthemum* require to be chilled but these are short-day plants. They survive from one year to the next by producing rhizomes just beneath the surface of the soil and these are vernalised by natural winter chilling. Shoots grow out from these rhizomes in the spring. When will these shoots flower? Explain your answer.

We saw in the above SAQ that chrysanthemums flower naturally in Autumn. Our knowledge of the control of flowering, however, coupled with the growth of plants in artificial environments, allows us to produce chrysanthemums in flower all the year round.

7.5.9 Why do fruit trees flower in Spring?

Having explained why chrysanthemums do not flower in spring, how do we explain the profusion of spring-flowering plants such as the flowering fruit trees? The answer here is that, although the flowers open in spring they were actually initiated and formed during the short days at the end of the previous year and passed the winter as dormant flower buds. Interestingly, these buds require to be exposed to winter chilling or they will not open. The same is true of the dormant vegetative buds which produce only leaves. You may have noticed that, in many flowering trees, the flower buds open before the leaf buds. This is due to their different sensitivities to temperature.

Π   We have studied three developmental phenomena in which chilling is a requirement. State all three and indicate which of them qualify to be called vernalisation.

The three are:

- the opening of flower buds of fruit trees;

- the cold treatment which some species need before they can respond to photoperiod and flower;

- the cold treatment which the seeds of some species need for their seeds to germinate.

Only the second is vernalisation. Be sure to keep this clear in your mind. It is often confused not only by students but also by eminent authors writing outside their area of expertise!

7.5.10 Plants also show a juvenile phase

In this, the final topic of the chapter, we will briefly discuss the phenomenon of juvenility in plants. Generally speaking, plants do not flower until they have reached a certain size or age irrespective of their exposure to photoperiod. Plants are said to be in the juvenile phase in this state and the length of it varies enormously. It is shortest in the Japanese Morning Glory, which can respond to short days as soon as its cotyledonary leaves have expanded. Most herbaceous plants cannot respond at such a young age and will react to photo-inductive cycles only when they have produced several leaves. The extreme cases are those of trees. For example, oak trees (*Quercus* spp.) remain in the juvenile (non-reproductive) phase for up to 30 years. The transition between the juvenile and the adult form is the result of an internal control mechanism and does not appear to be affected by the natural environment. This is another of the phenomena which have frustrated plant physiologists, and especially plant breeders, for many years. We know next to nothing about how it is controlled and can do very little to effect the change ourselves. A small amount of success has been achieved by Pharis' group in Canada. These scientists have discovered that a small number of coniferous species will flower precociously if they are injected with a mixture of gibberellins. However, the number of successful species is very small and in none of them is the transformation complete. That is to say, the trees require to be treated each year, they have not been permanently converted to the adult form. That having been said, this is a considerable achievement which has increased the rate at which the breeding of these species can be carried out.

Looking at flowering physiology as a whole, it is obvious in attempting to achieve our aim of controlling the flowering process considerable progress has been made but that there is still much to do. We are now ready to embark on the final chapter, to examine how plant growth itself is controlled.

Summary and objectives

Plants reproduce by asexual methods, which produce identical genetic copies of the parent and by sexual methods, which combine characters from two parents. Asexual methods include the production of bulbs, corms, runners, rhizomes and stolons. Sexual methods involve the formation of the flower which, after pollination and fertilisation, leads to the production of the seed. Most seeds are dormant when shed, further development being delayed. Seeds show epigeal or hypogeal germination, the aim of both being the establishment of the new seedling and the start of the next generation.

Flowering in many species is governed by photoperiod and plants can be categorised as SDPs, LDPs or DNPs. In SDPs and LDPs, the length of the night period is the critical time. The time-measuring device involves the pigment phytochrome which, from studies of germination in lettuce, is known to exist in two interconvertible forms, P_{730} and P_{660}. The evidence suggests that P_{730} at night is inhibitory to flowering in SDPs but stimulatory in LDPs and this led to the hour-glass model for time measurement. Recent measurements of the kinetics of phytochrome interconversion are not consistant with this model. The control of flowering may involve a flowering hormone, florigen, which appears to be universal in its operation but which has so far defied all efforts to isolate it. It is possible that florigen is a complex which includes one or more of the growth hormones. Biennials and some perennials need to be chilled before they can respond to photo-inductive cycles and this is called vernalization. Finally, all plants show a juvenile stage in their life when they are unable to respond to photoperiod. Forest trees show the extreme of this condition, which is a serious hindrance to the breeding of improved varieties.

Now you have completed this chapter you should be able to:

*   discuss the fundamental differences between sexual and asexual reproduction and show an understanding of several mechanisms for achieving the latter;

*   describe the development of seeds and fruits and point out the genetic origin of their parts;

*   list the factors which commonly effect seed dormancy and interpret experiments relating to this topic;

*   describe the difference between SDPs, LDPs and DNPs and interpret experiments on the effects of photoperiod on flowering;

*   describe the interconversion of P_{660} and P_{730}, propose a hypothesis to describe the involvement of this in time-measurement and describe experiments to test the hypothesis;

*   describe evidence which suggests the operation of a universal flowering hormone;

*   describe three developmental affects of chilling.

Growth and Development

Growth and Development

8.1 Introduction

In this, our final chapter, we will examine the effects of plant hormones on certain aspects of growth and development. Clearly, there is space to cover only a small part of this topic and our selection has been made to give you an understanding of some of the fundamental processes of plant growth and development, building on topics raised in earlier chapters, and to describe the role of hormones in these processes.

8.2 Plant growth occurs in meristems

primary meristems

Meristems are zones in which cell division is occurring and they invariably have associated with them zones of cell enlargement and differentiation. Primary meristems are found at the apices of stems and roots, and are responsible for growth in length of the stem-root axis. The organisational structure produced by these meristems is called the primary structure.

secondary meristems

Secondary meristems are responsible for growth in diameter and are called cambia. The vascular cambium produces new xylem and phloem, and the cork cambium produces cork. As a result of secondary meristematic activity, the primary structure is modified very significantly to produce what is called the secondary structure. Only dicotyledonous species produce secondary meristems. Monocotyledonous species do increase the diameter of their roots and shoots but by a different mechanism, as we will see later.

floral meristems

The stem apical meristem may, at some time, be converted from the vegetative to the floral state, in which case it produces sepals, petals, stamens and ovaries instead of leaves. Although the floral meristem is formed secondarily from the vegetative one, it is not spoken of as being primary or secondary.

8.2.1 The root apical meristem

Figure 8.1 shows the developmental activities of a root apical meristem.

The zone of cell division is not at the extreme end of the root because this is where the root cap is. Cell division in the cell division zone produces new cells which are produced either on the root cap side at A, (see Figure 8.1), in which case they become root cap cells, or on the other side, at B in the figure in which case they become part of the root axis. As each root cap cell dies, it releases mucilage, which is similar in chemical structure to the material we produce on our nasal membranes. Dying cells are replaced by new ones produced by the cell division zone so the root cap persists as a permanent entity.

| SAQ 8.1 | What is the function of the root cap? |

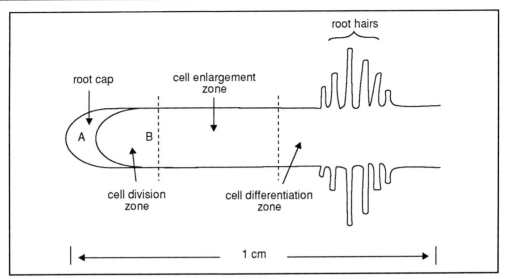

Figure 8.1 Diagram of a root apical meristem to show the main developmental zones. A and B are referred to in the text.

On the root axis side of the division zone there is first a cell enlargement zone and then a cell differentiation zone, but the transition between them is not as clear cut as Figure 8.1 might suggest. This is because not all of the cells in a zone are behaving in the same way. This is particularly the case with phloem cells. These cells are the first to begin to enlarge and then to differentiate and they can be recognised half way along the elongation zone. However, the main activity shown by cells in the three zones is that shown in Figure 8.1. Cells in the meristem and elongation zone possess only a middle lamella and primary wall. The secondary wall is laid down when growth ceases.

∏ How long is the cell division zone in the root shown in Figure 8.1?

The zone is approximately 1.5mm long. Species vary in this parameter but 1.5mm is a reasonable average.

Notice that there are no lateral roots indicated on Figure 8.1 which is because their formation occurs 1.5-2.0 cm behind the apex. Lateral roots are formed by the activities of certain cells of the pericycle, the layer of cells immediately inside the endodermis (Figure 8.2a).

lateral roots initiated by pericycle cells Lateral roots are initiated by pericycle cells at the poles of the primary xylem. These cells can be looked upon as potentially meristematic cells. When first formed they function only to deliver minerals to the xylem, as described in Chapter 4. Subsequently cells at the xylem poles divide several times to produce a zone of very small, densely cytoplasmic cells which becomes a replica of the root apical meristem. The lateral root meristems produce new cells behind their own apices and, as a result, grow in length and eventually grow out through the cortex and epidermis (Figure 8.2b) to become visible externally.

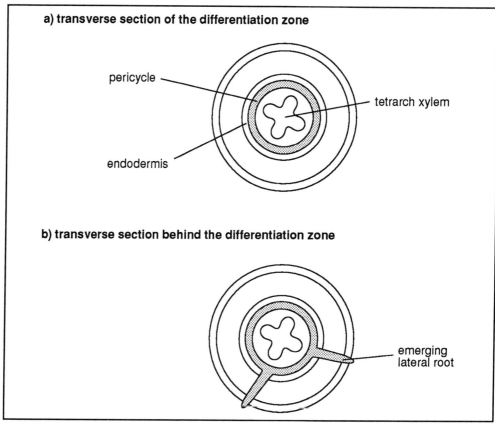

Figure 8.2 a) Transverse section of a root in the differentiation zone showing tetrarch xylem and the position of the pericycle; b) transverse section of a root behind the differentiation zone, showing emerging lateral roots.

8.2.2 The shoot apical meristem is more complex than the root apical meristem

Zones of division, enlargement and differentiation can be recognised in the shoot apex but the shoot apical meristem produces leaves and axillary buds in addition to the axial material of the stem, and is, therefore, more complex than the root apical meristem. New structures are formed in a repetitive manner by the shoot apical meristem and, as a result, the shoot is said to have a modular structure, illustrated in Figure 8.3, where each module consists of a leaf, its node and internode and the axillary bud.

shoot modular
structure

Cell division occurs in the apical dome but not at a uniform rate throughout. Cells towards the periphery of the dome divide more rapidly than those in the centre, giving rise to a bump or swelling on the flank of the dome. Continued cell division in the bump gives rise to a leaf primordium which eventually grows out to produce the leaf itself. Very soon after the initiation of the leaf primordium, a zone becomes demarcated at its base which develops into an axillary bud. When fully formed the axillary bud is a replica of the shoot apical bud in miniature. However, as we will see later, axillary bud development is arrested at this stage and recommences later in the life of the shoot.

leaf primordium

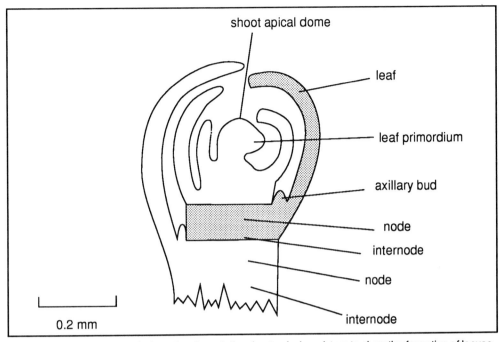

Figure 8.3 Diagram of a vertical section through the shoot apical meristem to show the formation of leaves and axillary buds and to illustrate the modular construction. The shaded zone demarcates a shoot module.

SAQ 8.2	Figure 8.3 shows that the shoot has no equivalent to the root cap. Is the apical dome protected in any way? If so, how?

activity of the shoot meristem is highly ordered

Note that the internode is formed as part of a module. Its subsequent development produces the major portion of the stem. The node and internode are initiated at the same time as the leaf primordium and in *Silene* there appears to be four layers of cells associated with each leaf, two giving rise to nodal cells and two to internodal cells. The situation in *Sambucus racemosa* is even more striking. This species produces tannin only in internode cells and these can be traced back to a single layer in the apical dome. In *Sambucus*, therefore, the node and internode each appear to be initiated as a single layer. These observations suggest that the activity of the shoot apical meristem is very ordered.

8.2.3 Formation of the internode involves further cell division

We have so far seen how the shoot module is formed. We will now examine how the internode grows to its final length. Figure 8.4 shows the results of observations designed to illustrate the occurrence and location of cell divisions in young stem.

The section used to produce Figure 8.4 was treated to show mitotic figures, *ie* cells containing chromosomes progressing through mitosis. Thus, each dot shows a dividing cells.

Π   Using the scale shown in Figure 8.4 decide how long the cell division zone is in the young shoot. How does this compare with the root in Figure 8.1?

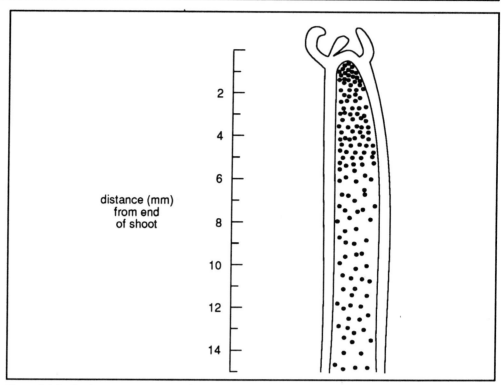

Figure 8.4 Stylised representation of distribution of mitotic figures in a young stem of *Chrysanthemum morifolium*. Each dot represents one mitotic figure in a 60μm thick longitudinal slice through the pith.

Mitotic figures are still evident more than 14mm below the apex in the shoot whereas division is complete within 1.0-1.5mm of the apex in the root.

The involvement of cell division in the formation of the full internode can easily be confirmed by counting cell numbers in vertical files of cells in sections of plants. Such sections can also be used to measure cell sizes. These analyses show that both cell division and cell enlargement occur as the internode grows.

The cell division activity in Figure 8.4 is occurring in the zone below the position of the meristem at the extreme apex of the stem. To distinguish the two, the meristem located in the apical dome is referred to as the eumeristem (true meristem) and the term sub-apical meristem has been coined for the lower, extensive meristem. The two together are referred to as the shoot apical meristem. The abbreviation SAM is used for the sub-apical meristem but care must be taken because S-adenosyl methionine is abbreviated in the same way.

eumeristem and sub-apical meristem

Finally, the enlarged cells differentiate to produce the cell types which we described in Chapter 1.

SAQ 8.3

Why is it proper to regard shoot meristems as being more complex than root meristems?

8.2.4 Shoots show tall, dwarf and rosette forms

Plants can be divided into two groups with regard to the growth form of their shoot:

- caulescent plants, which have an obvious stem with leaves separated by well developed internodes (for example, chrysanthemum);

- rosette plants, where leaves are borne on a very short stem in which internodes cannot be easily distinguished (for example, radish).

Caulescent plants can be further divided into tall and dwarf varieties, as we noted in Chapter 7, which differ in the length of their internodes.

How can we explain these differences in stem growth form? All three types produce leaves, nodes and internodes so the activity of their eumeristems, does not appear to vary. They do however, produce varying lengths of internodes so it is possible that their SAM activity varies. Is there any evidence for this? Examine Figure 8.5 and answer SAQ 8.4.

SAQ 8.4	If we assume that stem growth in *Samolus parviflorus* is controlled by the eumeristem and the SAM, which of the two has been activated by treatment with GA$_3$? Give reasons for your answers.

The data in Figure 8.5 suggest that the rosette species in question has an inactive SAM and that GA$_3$ can activate it. Data in agreement with this have now been obtained for a wide variety of species.

When dwarf varieties are examined for SAM activity numerous mitotic figures are seen, but these are increased significantly by treatment with GA$_3$, which stimulates internode extension in these varieties. This could suggest that dwarf varieties have partially active SAMs, which can be further activated by GA treatment. Taking our argument to its logical conclusion, tall varieties could be said to have fully-active SAMs. This is difficult to test because the plants already show active internode growth.

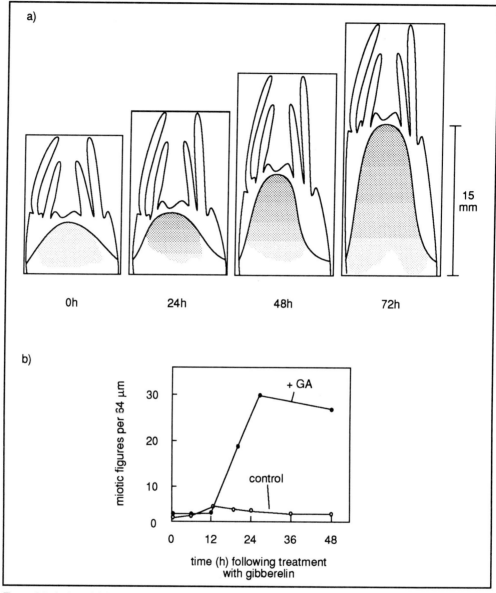

Figure 8.5 Action of GA₃ on stem growth in the rosette plant *Samolus parviflorus*. a) Shaded area represents cells undergoing mitosis following GA₃ addition; b) numbers of mitotic figures in 64 μm slices through the pith in plants treated with gibberelins. The control has not been treated with gibberelin.

SAQ 8.5

Can you think of a way to test the hypothesis? This is a tricky question and there is a hint in the response should you need it.

The results of experiments of the type described in the response to SAQ 8.5 are shown in Figure 8.6.

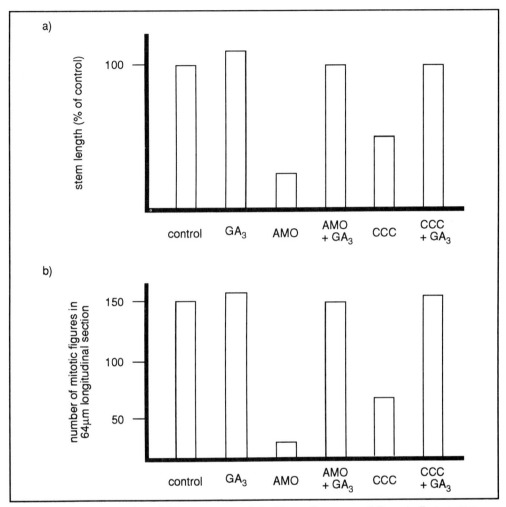

Figure 8.6 Effect of anti-GAs and GA3 on stem growth in *Chrysanthemum morifolium*: a) effect on stem length (percent of control), b) effect on mitotic activity in the apical 1.4 cm of the stem. AMO and CCC are examples of compounds which inihibit GA biosynthesis.

The results of these experiments are in agreement with our hypothesis that the activity of the SAM is an important controlling influence in tall as well as dwarf varieties.

In the experiments described above, the plant examples have all been dicotyledonous species. Stem growth in most monocotyledonous species is less easy to study because of the growth form of these plants before they flower. This is illustrated in Figure 8.7.

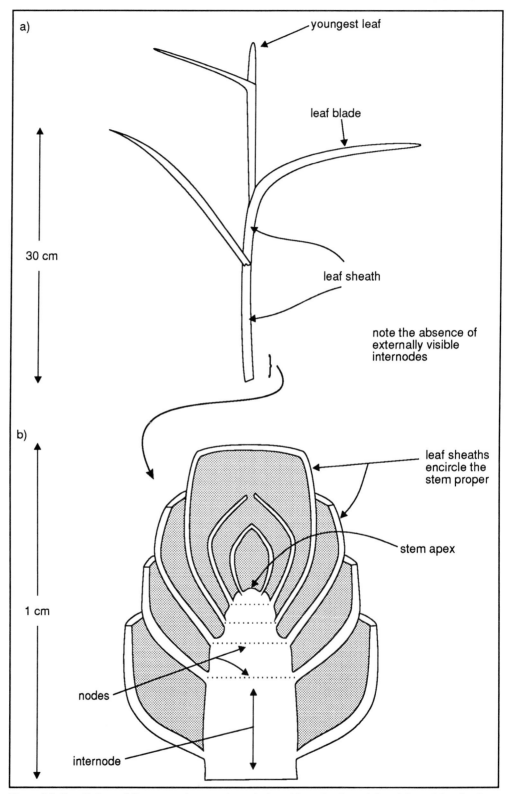

Figure 8.7 Growth form of *Zea mays*; a) external morphology, b) section through basal 1cm of the shoot.

During the early part of the growth of many monocotyledonous species the shoot consists mainly of leaf sheaths and blades, the stem itself being very short. We would describe this as a rosette form and in some species this form persists throughout the life of the plant. In others, such as *Zea mays*, stem extension occurs by the process of division and enlargement in the young internodes. Eventually, the internodes become visible externally by growing longer than the leaf sheaths encircling them. We would call these types of species caulescent. We noted earlier, that both tall and dwarf varieties of monocotyledons occur and experiments show that GAs can cause dwarfs to grow tall, that anti-GAs can cause talls to be dwarfed and that both division and enlargement of cells occurs during internode extension. For these reasons, stem extension growth is considered to be brought about in the same way in both monocotyledons and dicotyledons. The ability of CCC to reduce stem growth in wheat is utilised commercially to reduce the incidence of lodging (the blowing over of the cereal plants by rain and wind, a commonly observed phenomenon).

rosette forms

caulescent forms

lodging

8.2.5 Auxins and GAs interact in controlling stem extension growth

In the previous section we noted that GAs can cause cell enlargement in dwarf species and in tall species treated with anti-GAs. We also learnt in Chapter 6 that IAA can cause cell elongation.

∏ Is the action of these two hormones the same?

The answer is no, because IAA does not cause dwarfs to grow tall and GAs are not active in the oat coleoptile test, one of the bioassays for auxins. Generally speaking there are very few instances in which the applications of natural auxins to intact plants leads to any increase in elongation; whereas we know of many such instances with GAs. In contrast, auxins often cause the elongation of excised segments of tissue whereas GAs do not. Bearing in mind that a hormone will cause elongation only if there is a deficiency of that hormone in the tissue, these data are interpreted as reflecting a differing hormone status in the tissues. A tissue which does react to both auxins and GAs is the excised young internode of light-grown peas. Experiments show that 5×10^{-5} mol l^{-1} IAA and 10^{-4} mol l^{-1} GA$_3$ both produce optimal extension of these pea segments. Interesting results were, however, obtained when the two hormones were applied at the same time (Figure 8.8).

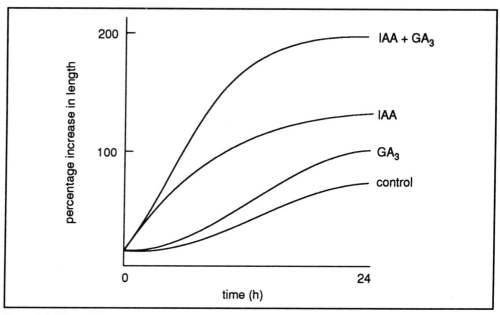

Figure 8.8 Effect of IAA and GA₅ added separately and together on extension growth of green pea stem sections (based on the data from R.Ockerse and A.W. Galston (1967), Plant Physiology 42: 47-54).

SAQ 8.6

Figure 8.8 shows two reasons why the action of GAs and IAA on growth is separate and distinct. Try to think what they are and give your reasons.

The synergistic response of pea stem sections shows that GAs and IAA act on separate but related aspects of the system which controls growth. We now need to examine growth at the cellular level to see if we can explain their action.

8.3 Cell growth involves the formation of the vacuole

Cells in a meristem have a diameter of 10-15 μm. This diameter seems to be equal in all directions ie isodiametric. If we assume that this is similar to a sphere we can calculate its volume from $4/3 \pi r^3$. Mature cortical cells are cylindrical in shape and are of the order of 100 μm long by 25 μm wide. The volume here can be calculated from $\pi r^2 h$.

∏ Calculate the approximate volume of a meristematic cell and a mature cortical cell, assuming the meristematic cell to be 10μm in diameter.

The meristematic cell volume is of the order of 520μm³, while the cortical cell volume is approaching 50 000 μm³.

This calculation illustrates that cells show an almost 100 fold increase in volume as they grow. Most of this increase in volume is water which occupies the vacuole. We will now examine the process of formation of the vacuole (Figure 8.9).

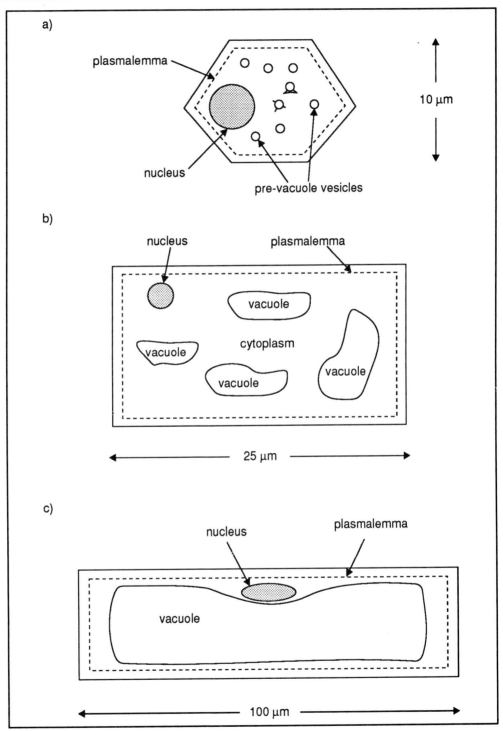

Figure 8.9 Stages in the formation of the vacuole (see text for details). Note that organelles other than the nucleus and the vacuoles have been omitted and that the scale changes from a) through b) to c).

The meristematic cell in Figure 8.9a is characterised by what is called a dense cytoplasm, because of the number and proximity of mitochondria, Golgi bodies, endoplasmic reticulum and ribosomes. It also contains a considerable number of colourless vesicles which have been derived either from Golgi vesicles or portions of the endoplasmic reticulum or both. A cell which has embarked on enlargement (Figure 8.6b) can be seen to contain a small number of medium-sized vacuoles whereas the mature cell (Figure 8.9c) has one large one. The implication of such a series of observations is that the small vesicles in the meristematic cell are the vacuole precursors, that they absorb water and swell and coalesce into gradually larger and larger entities until there is just one large vacuole.

8.3.1 Cell growth involves water absorption

The rate of cell enlargement ultimately derives from the balance between the yielding properties and extensibility of the cell wall and the ability of a cell to take up water from its surroundings. You should remember the water potential equation in which:

$$\psi = \pi + P$$

For a cell to absorb water, its ψ must be below that of its surroundings. If water is absorbed, the cell volume increases and pressure is exerted on the cell wall. In a mature cell, the cell's turgor pressure increases until it counteracts the effect of solutes and no more water absorption occurs. In a young immature cell, the cell wall yields to the turgor pressure and the cell grows. Figure 8.10 shows this relationship.

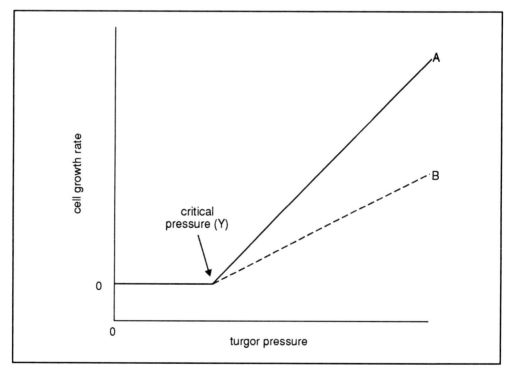

Figure 8.10 Relation between turgor pressure and growth rate. A and B are referred to in the text. The critical pressure is the turgor pressure that has to be reached before cell growth occurs. Critical pressure (Y) is also referred to as the threshold pressure or the wall yield stress (see text).

critical
(threshold)
pressure
wall
extensibility

hydraulic
conductance

wall yield
stress - critical
pressure

Note that cell growth rate does not increase until turgor pressure has reached a certain value. This is referred to as the critical or threshold pressure above which growth rate is proportional to pressure. Note that Figure 8.10 shows growth rates for two tissues, A and B, with identical critical pressures. These differ in a wall property referred to as wall extensibility. Tissue A has a higher wall extensibility than tissue B, the actual value being calculable from the slope of the line. In addition to the factors already mentioned, the rate of growth depends upon the hydraulic conductance, which is a measure of the ability of water to pass through cell membranes. We can represent this relationship in the following way:

$$\text{growth rate} = \frac{L_p W_{ext}[(\psi_{ext} - \psi_{int}) - Y]}{L_p + W_{ext}} \tag{E-8.1}$$

where L_p is the hydraulic conductance through the tissue, W_{ext} is the wall extensibility, ψ_{ext} and ψ_{int} are external and internal water potential respectively. Y is the wall yield stress, which is another name for the critical pressure.

∏ Let us examine this equation in a little more detail. Does it apply at all values of ψ_{ext} and ψ_{int}? (Figure 8.10 might give you a clue).

The answer is no. What the equation effectively says is that the rate of growth is proportional to $[(\psi_{ext} - \psi_{int}) - Y]$. The difference between ψ_{ext} and ψ_{int} effectively leads to water movement into the cell and the generation of turgor pressure. Providing the turgor pressure is greater than Y, then growth will occur (see Figure 8.10). If, however, it is equal to, or less than Y, then no growth will be observed (see Figure 8.10). For turgor pressure above Y, growth rate is proportional to the turgor pressure minus Y.

In other words our equation holds provided $(\psi_{ext} - \psi_{int})$ is equal to or greater than Y. If $(\psi_{ext} - \psi_{int})$ is smaller than Y, then the growth rate will be 0.

SAQ 8.7

Answer the following questions about Equation 8.1.

What would happen to the growth rate if:

1) internal and external ψ were identical?

2) L_p was increased?

3) W_{ext} was decreased?

4) Y was decreased?

5) ψ_{int} was decreased, ie made more negative by an increase in the solute concentration of the cell?

8.3.2 Cells show plastic and elastic extensibility

Analyses of the parameters in Equation 8.1 have been carried out by numerous scientists. The results show that L_p is relatively stable while cells are growing and so does not contribute to changes in growth rate. Similarly, there is no evidence for an increase in the solute concentration of growing cells so a change in solute concentration does not appear to be the driving force. On the other hand, there is good evidence for changes in wall properties as a cell is growing.

Plant tissues exhibit both plastic and elastic extensibility. This can be demonstrated by the use of an Intron analyser which can be used to apply weights to a plant tissue and to determine the effect (Figure 8.11).

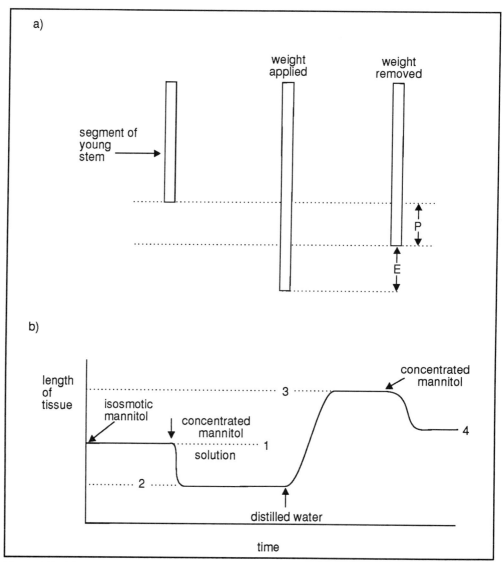

Figure 8.11   Effect of application and removal of a) a weight to a segment of young stem: b) mannitol to a segment of young stem. E, P and numbers are referred to in the text.

When the weight is applied, the tissue stretches but rebounds a little when the weight is removed. The amount of rebounding, E in Figure 8.11a, is a measure of the tissue's elastic property. The tissue is still longer than its original length since it has been stretched. This irreversible stretching (P in the figure), is a measure of the plastic property of the tissue. Figure 8.11b shows that similar phenomena can be observed in young pea stem segments when exposed to mannitol, a non-absorbed osmotically-active compound. Tissues in a concentrated mannitol solution become plasmolysed, ie turgor equals zero and the tissue shrinks. When placed in water, the ψ

is very high, turgor increases and the tissue expands. When placed in mannitol again, it shrinks by the same amount as before. Total expansion is represented by the difference in length between 2 and 3. This has two components elastic extension (1-2 or 3-4) and plastic extension (1-3 or 2-4).

Numerous pieces of evidence show that the plastic extensibility of tissues increases when cells are stimulated to grow (Figure 8.12).

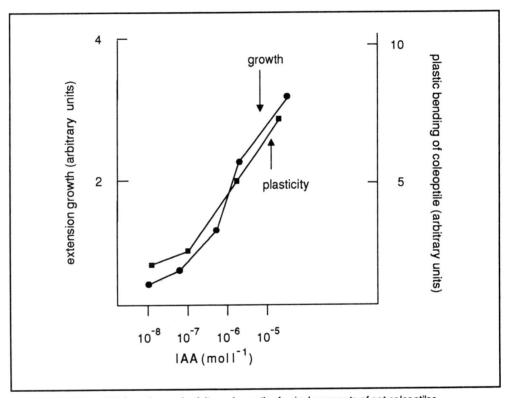

Figure 8.12   Effect of IAA on tissue plasticity and growth of apical segments of oat coleoptiles.

Gibberelins
and auxins
increase
plasticity but
not elasticity

Figure 8.12 shows a good correlation between growth of the coleoptile and the plasticity of the tissue. It further suggests that IAA acts by causing an increase in plasticity. Other tests show, however, that IAA does not enhance elasticity.

Sections of cucumber hypocotyls are a good tissue to demonstrate the action of GAs. GAs stimulate growth. Again plasticity is increased by GA treatment but elasticity is not affected.

Figure 8.13 shows the results of further experiments with IAA and GAs.

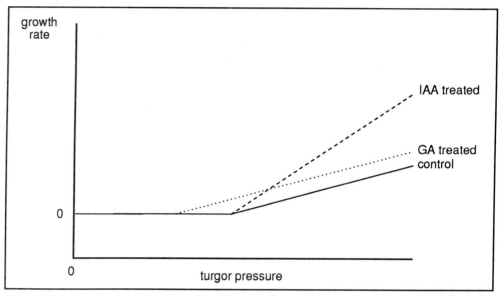

Figure 8.13 Effect of turgor pressure on the growth rate of IAA-treated and GA-treated plant segments.

SAQ 8.8   What parameter(s) in Equation 8.1 has(have) been altered by treatment with IAA and with GA, as shown by Figure 8.13?

∏   Draw a line on to Figure 8.13 which would show the effect on growth rate if IAA and GA were applied together.

The answer is shown in Figure 8.14.

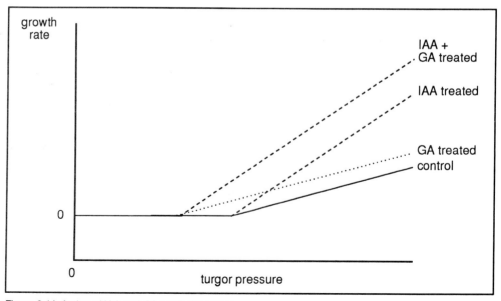

Figure 8.14 Action of IAA and GA applied together.

Wall extensibility and wall yield stress are both aspects of the plastic property of the wall. This knowledge of the actions of IAA and GA explains their synergistic action, shown in Figure 8.8.

We now need to examine wall structure to see if we can account for these wall properties in some aspect of wall architecture.

8.3.3 The cell wall consists of a set of interlinked networks

Cell wall structure has been studied in two basic ways. The original method involved the separation of components by their solubility properties followed by the chemical characterisation of these components in Figure 8.15. While this has provided much useful analytical information it provided no details of how the components are attached to each other. Such information has been provided by the second, more recent approach, that of breaking cell walls into pieces using fungal and bacterial digestive enzymes, followed by the characterisation of the pieces.

Figure 8.15 shows how wall components can be separated by solubility.

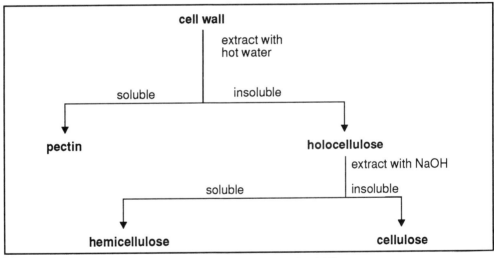

Figure 8.15 Separation of cell wall components according to their solubility.

Cell walls consist of two phases, a microfibrillar phase and a matrix phase. The microfibrils are made of cellulose, which differs from the matrix phase by a high degree of crystallinity and a relatively homogeneous composition. Cellulose microfibrils consist of chains of β 1,4-glucose residues of considerable but uncertain length. In some tissues there is close to 15,000 glucose residues in each polymer. The cellulose fibre contains 30-100 poly-glucose molecules lying side by side forming a fibre about 10 nm in diameter. These are large enough to be visible in the electron microscope.

matrix phase contains pectins and hemicellulose and extensin

The matrix phase appears relatively featureless in the electron microscope but is, in fact, extremely complex. Pectin and hemicellulose contribute to the matrix and there are at least six forms of each present. In addition, a structural protein called extensin is present. The composition of the matrix varies in different parts of the wall, in different types of cell and in different species. The components are not easy to analyse chemically and hence we are still some way from obtaining the complete picture. Figure 8.16 gives a simplified view which will form the basis of our discussion. It is quite complex so take a little time over examining it. You will see that we have drawn cellulose microfibres and shown a number of other types of molecules associated with these cellulose microfibres. We will explain a little more about the interactions between molecules in the following paragraphs and illustrate some of these interactions in Figures 8.17 and 8.18.

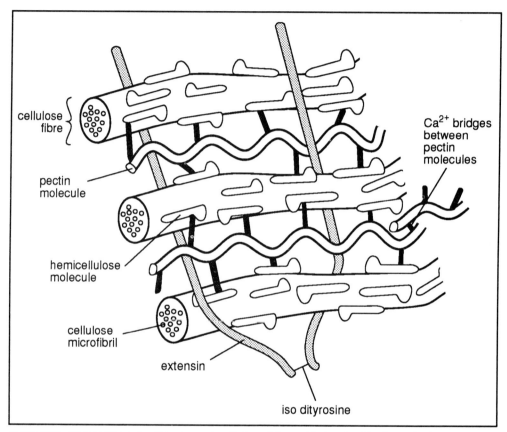

Figure 8.16   Model of the cell wall to show microfibrils and the matrix (see text for discussion).

The cellulose fibres in the wall are partially coated with hemicellulose as shown in Figures 8.16 and 8.17 a).

Note that there are many hydrogen bonds between the various molecules. Although there are several different hemicelluloses, a major one is a mixed polymer of xylose and glucose called xyloglucan.

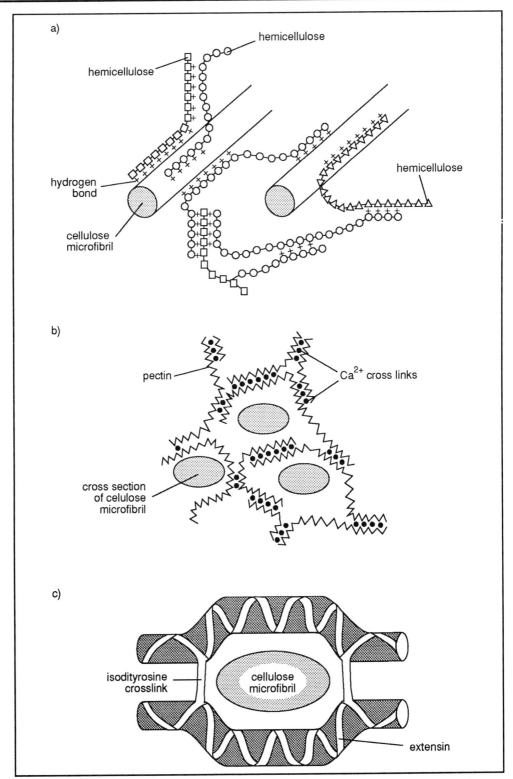

Figure 8.17 Stylised figures showing a) interlinking between hemicellusoe and cellulose, b) interlinking a pectin molecules by calcium bridges and c) interlinking of extensin by isodityrosine crosslinks (see Figure 8.18).

Hemicellulose molecules are covalently linked to pectin molecules, which consist of long polymers containing many residues of galacturonic acid. Adjacent pectin molecules can be linked by calcium bridges which bind to the carboxyl groups of the galacturonic acid residues. This can form a considerable measure of interlinking (Figure 8.17 b).

Also running between the cellulose microfibrils is the protein extensin which can form cross links between tyrosine residues (Figure 8.17c and 8.18).

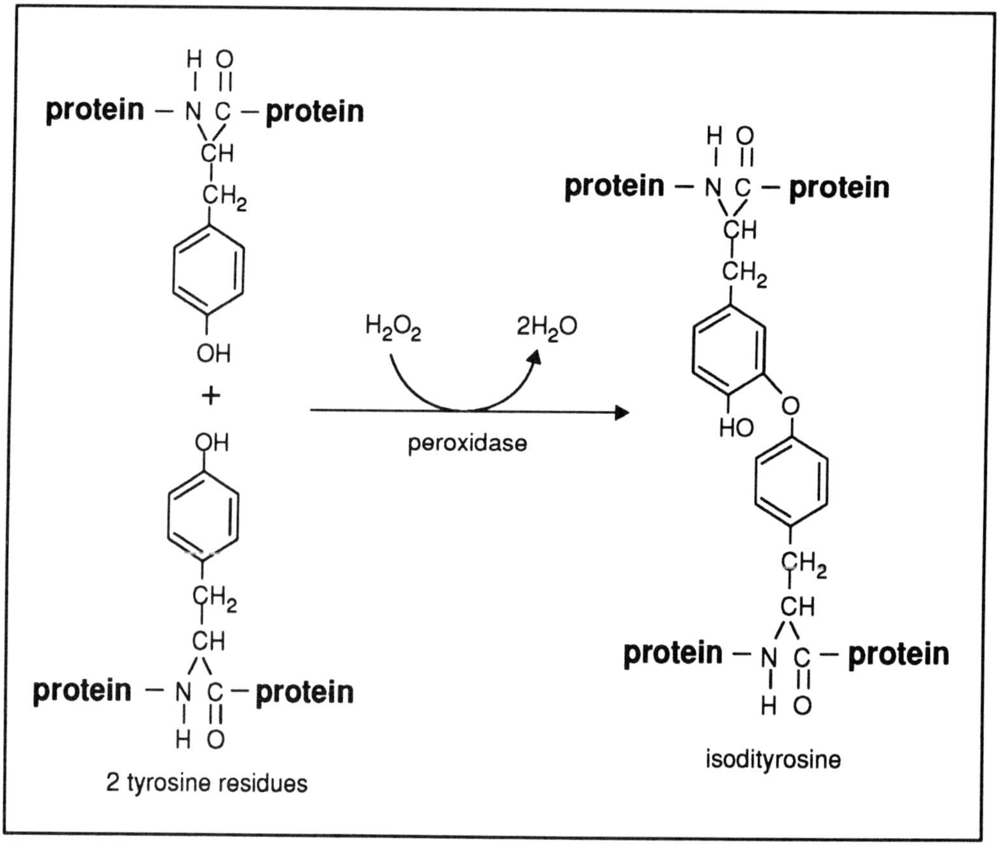

Figure 8.18 Formation of isodityrosine.

If you now go back to Figure 8.16, you will see that it is indeed a simplified diagram. Nevertheless, it will serve as a useful basis for discussion of the effects of auxin and GAs. The nature of the wall, ie the fact that it consists of largely crystalline microfibrils embedded in an amorphorous matrix, means that it will have natural elasticity. If the microfibrils were directly and rigidly linked to each other the wall would show little or no elasticity. Several models have been proposed to explain plasticity. Figure 8.19 shows one.

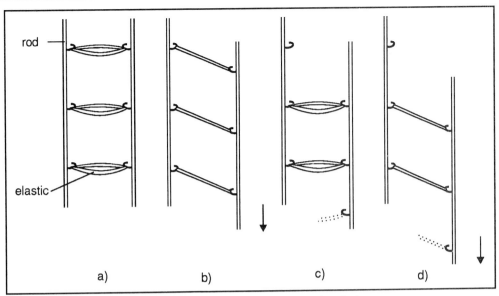

Figure 8.19 Model for plastic extension of cell walls (see text for explanation).

cell wall
plasticity may
be dependent
upon the
breaking and
reformation of
bonds

Imagine two rods linked by elastic bands (Figure 8.19a). As the rods are pulled in opposite directions the rubber bands stretch to their maximum (Figure 8.19b). However, if one end of each cross-linking band is unhooked and reattached (as shown in Figure 8.19c) the rods can be pulled apart again (Figure 8.19d). This model suggests that plastic extension involves the breaking and reformation of bonds between components in the wall. Observations of cellulose microfibrils in growing cells shows that they are oriented in a generally perpendicular direction to the long axis of the cell and that they are pulled apart and reoriented during growth (Figure 8.20).

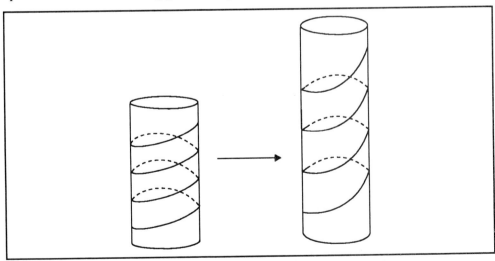

Figure 8.20 Changes in position and orientation of cellulose microfibrils during growth.

These observations are interpreted as showing that the breakage of bonds in the matrix allows the microfibrils to be pulled apart and that this constitutes the plasticity of the cell. Unfortunately, it does not allow us to distinguish between wall extensibility and the yield threshold. Let us now examine evidence for the effects of auxins and GAs on wall properties.

8.3.4 The acid growth hypothesis

It has been known since the 1930s that acid solutions can cause elongation growth of segments of coleoptiles and hypocotyls. This was rediscovered in the 1960s when it was shown that IAA caused coleoptiles to acidify the medium in which they were incubated. The addition of neutral buffers prevented the change in pH and prevented the growth stimulating action of IAA. Some scientists are still not convinced that auxin action is mediated through the production of acid. The hypothesis has, however, been greatly strengthened by some evidence reported in 1991. It has been possible to produce antibodies to the plasma membrane H^+-ATPase of *Zea* coleoptiles and to tag them with a fluorescent label. This enables H^+-ATPase to be measured quite accurately. It has been shown that auxin treatment of coleoptiles caused an increase in the amount of detectable H^+-ATPase in the plasma membrane. The kinetics of this were very similar to that of growth stimulation (that is, detectable after 10 mins and reaching a maximum after 30 mins). This phenomenon was studied further by the use of inhibitors of protein synthesis and of RNA synthesis. Both inhibitors significantly reduced auxin-induced growth and H^+-ATPase increase. These results are shown diagrammatically in Figure 8.21.

auxin action mediated by acid

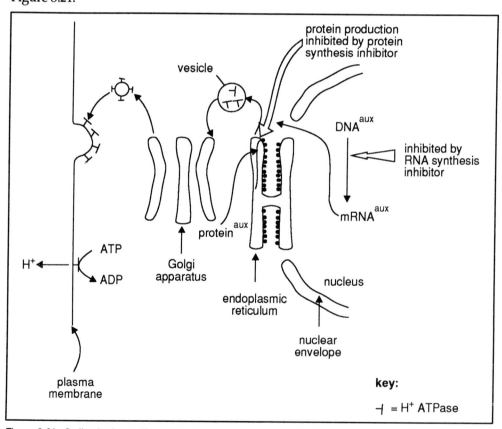

Figure 8.21 Stylised scheme illustrating auxin-induced wall acidification. To follow this scheme, begin at DNA^{aux}. Exposure of a cell to auxin is believed to induce the transcription of specific gene(s) (represented by DNA^{aux}) leading to the production of the appropriate mRNA(s) (represented by mRNA^{aux}). This is transported to the endoplasmic reticulum where it is translated into protein and passed into the lumen of the endoplasmic reticulum. Here the protein is packaged into vesicles which migrate to the Golgi apparatus where it is further processed and packaged onto Golgi-derived vesicles. These migrate to and fuse with the plasma membrane. At least some of the mRNAs produced in this scheme code for H^+-ATPase. Note stages at which inhibitors of protein synthesis and of RNA synthesis block the production of H^+-ATPase.

These data have brought the acid growth theory back into the centre of the picture and have resulted in the stimulation of further research activity in this area.

protons may activate wall loosening factor

The secreted protons are considered to activate a wall loosening factor. The nature of this is not known but it is suspected to be an enzyme which breaks one or more bonds in the wall matrix thus increasing plasticity. We do not yet know the nature of this enzyme but it is thought that its substrate is one or more of the hemicellulose molecules. This cannot be the whole picture regarding the action of IAA, however. We have seen that IAA treatment causes a growth increase which peaks after about 30 mins. In coleoptiles and hypocotyls of many species, the growth rate falls about one hour after treatment but then rises to another peak one to two hours later. The application of acid causes the early growth increase but not the latter, so proton secretion may be only part of the reaction to IAA.

GAs do not cause proton secretion although they too cause growth increase within 10 minutes. It has been shown, however, that GAs cause the rapid absorption of calcium ions. It is known from other work that the application of high levels of calcium ions inhibits GA-stimulated growth. The results of an experiment to test for the effects of calcium ions on plasticity are shown in Figure 8.22.

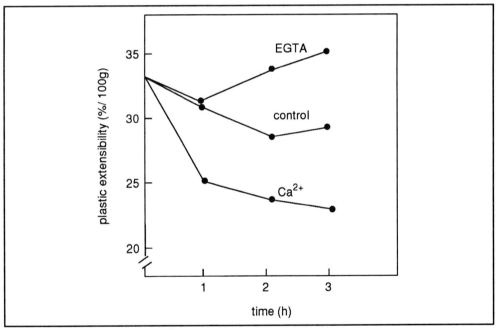

Figure 8.22   Effect of addition of Ca^{2+} or removal (EGTA) of calcium ions on plastic extensibility of soybean hypocotyls. (EGTA is a chelating agent which binds calcium ions).

Notice that in the control treatment plastic extensibility falls during the first 2 hours but then remains steady.

SAQ 8.9

What are the results of the Ca^{2+} and EGTA treatments?

Ca²⁺ in
apoplast
influence
cross-linking of
pectins

These results raise the possibility that the action of GAs on growth is mediated by the concentration of calcium ions in the apoplast. It is tempting to suggest that this modifies the degree of cross-linking of pectins (Figure 8.17b), thereby increasing plasticity. The consensus is, however, that this may be too simple an explanation.

You can see from this last section that much progress has been made in understanding the control of plant cell growth but we are still someway from a complete understanding. Such an understanding is necessary if we are to be able to manipulate growth to our benefit.

8.3.5 Do solutes play any part in cell growth?

In our discussion so far we have noted that water movement into a growing cell could theoretically be caused by an increase in the cell's solute concentration or by changes in wall properties which allow turgor pressure to be more effective. We noted that there is no evidence for an increase in solute concentration during the growth process and we have concentrated our attention on changes in properties of the wall. However, this must be an incomplete picture. Let us consider why.

Re-examine Figure 8.9a showing a meristematic cell. Imagine that the cell's ψ is lower than the solution's and that the yield threshold and wall extensibility have now been changed, let us say, by application of IAA and GA_3. Now attempt to answer the following SAQ.

SAQ 8.10

Will there be any movement of water? If so, try to decide where it will be delivered to.

To reiterate the response to the SAQ, if we simply modify wall properties, the cytoplasm will become flooded with water.

Does this happen? Re-examine Figure 8.9 if you are unsure.

transport of
solutes into
pre-vacuolar
vesicles

The answer is, 'No!'. Water does not flood the cytoplasm, rather it is delivered to the small pre-vacuolar vesicles which enlarge and subsequently coalesce. How does the plant do this? No one knows the exact answer but it is suspected that the cell directs water to the pre-vacuolar vesicles by increasing their solute concentration, ie by transporting solutes into these vesicles. This initiating phenomenon must involve the redistribution of solutes within the cell for, as we saw earlier, overall cell solute concentration does not increase during cell enlargement. Notwithstanding the lack of an increase in solute concentration (expressed in terms of MPa), there is considerable evidence to suggest that solutes are very actively delivered to cells as they grow. Thus, despite the approximately hundred fold increase in volume as a cell grows, the solute concentration in pea epicotyls, for example, falls only from 0.9 to 0.8 MPa. Strikingly, as maize roots grow, their solute concentration shows no detectable fall. We do not know the nature of the solutes which are used to initiate vacuolation but, once growth has started, there is a direct link between growth and phloem transport. In pea epicotyls as much as about 80% of incoming sugars serve to provide osmotically-active solutes which are delivered to the vacuole.

There is one final point to make about the process of vacuolation, that of the problem of membrane recognition.

8.3.6 The problem of membrane recognition

The description of Figure 8.9 spoke of the coalescence of the growing vacuoles into one big one but this simple statement hides a very complicated phenomenon. It is not appropriate to spend a lot of time on this but it is important that you are aware of the problem, which can be stated in the form of a question. How do the pre-vacuolar and later vacuolar vesicles know which other vesicles to fuse with? They need a recognition system which allows them to identify each other and avoid coalescing with, for example, mitochondria or chloroplasts or even the nucleus. This recognition system is presumably inserted into the vesicle membrane as it is formed from the Golgi apparatus or endoplasmic reticulum, thus preventing it from re-fusing with its 'parents'. The recognition sites are presumably proteins in the membrane. Information about these is only slowly becoming available and, as yet, is far from complete.

A second problem concerns the act of coalescence itself which must involve a considerable degree of rearrangements of cellular constituents.

We hope you will agree that the process of vacuolation is a complex but fascinating subject.

Before we leave the topic of cell growth we should examine the effects of light on the process.

8.3.7 Light also affects cell growth

When we discussed phytochrome in Chapter 7 we noted that it exists in two interconvertible forms:

$$P_{660} \xrightleftharpoons[\text{far-red light or darkness}]{\text{white/red light}} P_{730}$$

photo-equilibrium (photostationary state of P_{660} and P_{730})

You may be under the impression that the interconversion is an all or nothing response, that all the phytochrome is in the P_{730} form in red light or all in the P_{660} form under far-red light. This is not the case. A re-examination of Figure 7.16 shows that there is considerable overlap of the absorption spectra of the two forms of phytochrome. Note, for example, that even at the absorption maximum of P_{660}(660 nm), P_{730} shows considerable absorbance. The same phenomenon occurs at the absorption maximum of P_{730}. The outcome is that neither of the phototransformations goes to completion and, under saturating conditions, the two forms are present in photo-equilibrium, sometimes referred to as photostationary states. The actual position of the equilibrium is governed by wavelength and irradiance (fluence rate). This equilibrium is represented by the Greek letter φ and is the amount of P_{730} as a proportion of P_{total}. Thus, red light (660 nm) produces a φ of 0.8 (ie 80% P_{730}), whereas far-red light (730 nm) produces a φ of 0.03 (ie 3% P_{730}). A very wide range of φ values between 3 and 80% can be brought about by using single or multiple wavelengths of light at various fluence rates. What happens when we grow plants under these different φ values? Typical results are shown in Figure 8.23 for an open site weed, *Chenopodium album*.

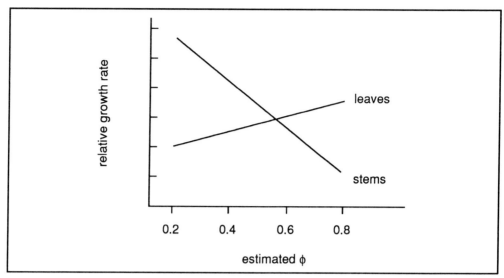

Figure 8.23 Effects of φ on leaf and stem growth rate in *Chenopodium album* (fat hen).

The results show that leaves and stems behave in an opposite manner. As the proportion of P_{730} falls, stems grow more quickly but leaves more slowly. The extreme of this situation is obtained when plants are grown in the dark. Under these conditions the level of P_{730} is zero. Here stems are very long and thin, leaves very small and the plant is said to be etiolated.

∏ Can you give one more symptom of plants grown in the dark?

They are very pale in colour because they do not synthesise chlorophyll.

To explain the significance of these phenomena we need to examine the spectral distributions of energy in various situations. The spectral distribution of full sunlight (Figure 8.24 a)) contains significant amounts of light energy between 400 and 800 nm.

Light which has passed through a canopy of leaves has a much lower amount of photosynthetically active radiation but, perhaps more important, a very different ratio of red to far-red light.

∏ Calculate the approximate red: far-red ratios in Figure 8.24 a) and b) using 660 nm as red and 730 nm as far-red. Then determine the values of φ which would be generated by these ratios from Figure 8.24 c).

The red: far-red ratio for full light is approximately 1.1 and for the canopy is approximately 0.2. The respective φ values are 0.58 and 0.25.

Examination of Figure 8.23 shows that these values support quite different stem and leaf growth rates. Relative to full sunlight, the canopy light causes increased stem growth but decreased leaf growth. The significance in these figures and growth activities is that it shows how a plant will react if it is in the shade of another plant. Put another way, the plant detects that it is shaded by another plant by perceiving the change in the red: far-red ratio of light falling in it.

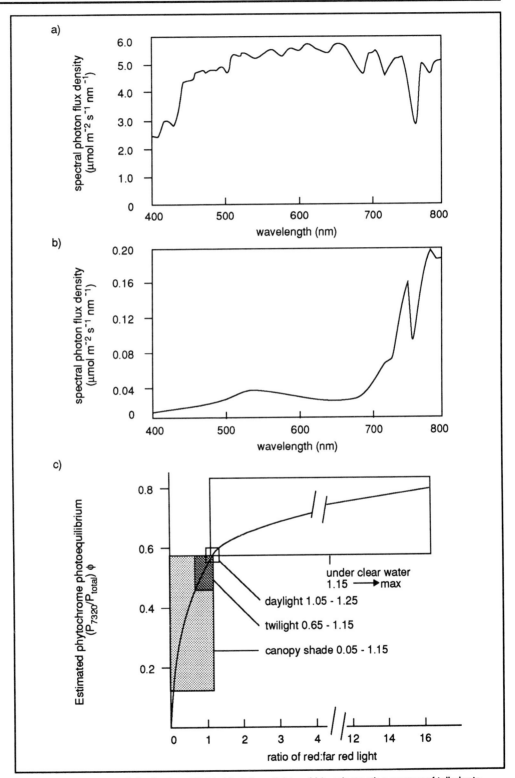

Figure 8.24 The spectral distribution of light a) in full sunlight and b) underneath a canopy of tall plants, and c) the effect of red: far-red ratio on φ.

<table>
<tr><td>

SAQ 8.11

</td><td>

A fat hen plant which is overshadowed by other plants will show an increased rate of growth of the stem but a decreased rate of leaf growth. What will happen regarding growth when the plant grows out from the shade of the surrounding plants?

</td></tr>
</table>

8.3.8 Some plants do not respond to shade

shade avoiders and shade tolerators

The examination of the response of a wide range of species to shade and to various ratios of red to far-red light reveals that some species do not react in the way we described in the previous section. Species can be divided broadly into shade avoiders and shade tolerators. *Chenopodium album* (fat hen), the species referred in Figure 8.23, is a shade avoider. In the shade it shows enhanced stem growth and reduced leaf growth, as indicated. In addition, petioles are longer and axillary bud outgrowth reduced, the plant thus showing increased apical dominance. *Chenopodium* is a competitive weed species normally found in open environments. Its reaction to the shade of other species could be said to confer a selective advantage. Many species, however, do not react in this way. Woodland plants, such as *Mercurialis perennis* (dog's mercury) normally grow in the shade of trees. These species tend to be slow growing and show very little response if transferred from the shade to an exposed position. Stem and leaf growth show a much reduced response to changes in the red to far-red ratio of light. They would obviously not be able to grow sufficiently to pierce the tree canopy above and thus appear to be well adapted to their environment. Plants which belong to the *Chenopodium* group are called shade-avoiding species while *Mercurialis* belongs to the group known as shade-tolerant species. Most crop species are, not surprisingly, shade avoiders.

We can now return to our topic of the factors affecting cell growth. A number of observations indicate that light-induced changes in stem growth rate are correlated with changes in wall plasticity. The increase in growth rate of etiolated stems is accompanied by an increase in plasticity as compared with plants grown in full light. Further, the exposure to the etiolating conditions causes the secretion of protons into the apoplast and the growth increase can be reduced with neutral buffers.

These results and those described earlier focus, attention on the central role of plasticity in the control of cell growth rate. The phenomenon of proton release raises the possibility that auxins and light act through one and the same receptor or at the same membrane location, but much more work needs to be done before we can be sure of this.

8.4 Vascular cambial activity controls growth in diameter

Growth in diameter in dicotyledonous species is brought about by the activity of the vascular cambium as shown in Figure 8.25.

development of the secondary structure of roots and shoots

In the primary stem and root, the vascular cambium is present between the primary xylem and phloem. The cell division activity of the cambium produces new cells either towards the inside or towards the outside. Cells which remain in the central zone retain the meristematic function but those produced on the inside or outside go through the differentiation process. The outer ones become phloem, the inner ones xylem. Eventually both the root and the shoot consist of a central solid core of xylem tissue with a continuous ring of phloem around it, separated by the vascular cambium. This new arrangement of tissues is referred to as the secondary structure.

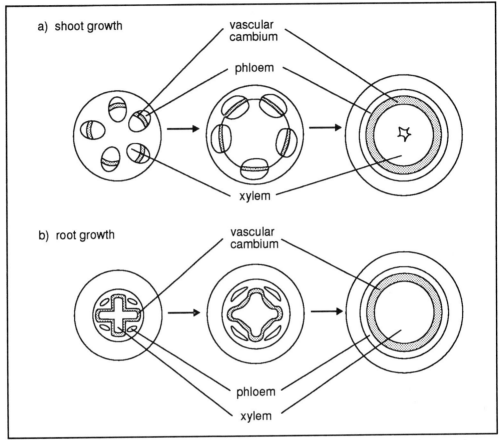

Figure 8.25 Diameter growth in dicotyledonous shoots (a) and roots (b). The figures on the left are of primary shoots and roots, whilst those on the right are of secondary shoots and roots (not to scale, see text for further discussion).

variable
composition of
swollen roots

The activity of the vascular cambium is of great commercial importance. An obvious example is the production of timber, but there are several others such as the swollen roots of sugar beet, beetroot, carrot and radish. There are very few lignified cells produced in plants which produce swollen edible roots. These species also vary with regard to the amount of xylem and phloem they produce. Radish swollen roots consist mainly of xylem parenchyma whereas carrot, beetroot and sugarbeet contain mainly phloem parenchyma.

Little work has been done to elucidate the role of plant hormones in the control of cambial activity in edible roots, but we do have some information from work on trees. This utilises the fact that in trees cambial activity is seasonal; it lies dormant during the winter and then begins to grow again in spring. In many species cambial activity begins immediately below expanding buds and spreads slowly downwards, ie in a basipetal direction, through the twigs to the branches and trunk and then into the roots. In the roots the spread of activity also occurs downwards but here, of course, the direction is an acropetal one.

SAQ 8.12

What hormone is transported in the same way as the spread of cambial activity? What is the name of the transport system?

We noted in Chapter 7 that auxins are synthesised in young leaves which raises the possibility that they are involved in the process of cambial reactivation. This idea is supported by the fact that cambial reactivation does not occur if the buds are removed from twigs, but does occur if IAA is applied to the upper end of the disbudded twig. A careful analysis of the activity of IAA, however, revealed that although new cells were produced by the cambium, these were mainly on the xylem side and very little phloem was produced. The examination of the action of a wide range of compounds revealed that GAs, were the only other hormones active in this system. GAs, however, caused mainly phloem to be produced. These results are shown in Table 8.1.

Treatment	Xylem width*	Phloem width*	Total width*
control	1	0	1
100 mg l^{-1} IAA	3	0	3
100 mg l^{-1} GA$_3$	1	9	10
100 mg l^{-1} each of IAA and GA$_3$	11	12	23

Table 8.1 Quantitative effects of IAA and GA$_3$ on cambial activity in disbudded twigs of poplar. *Values are relative to the amount of xylem produced by the control. (After P.F. Wareing, C. Hannay and J. Digby, in The Formation of Wood in Forest Tress, Academic Press, New York, pp 323-344, 1964).

SAQ 8.13

From the data given in Table 8.1, how would you describe the action of IAA and GA$_3$ when applied together?

auxin and GAs control xylem and phloem production by vascular cambium

These results are interpreted as showing that auxins and GAs combine with each other in the control of xylem and phloem production by the vascular cambium. We saw in an earlier section that these two hormones are involved in the control of extension growth and we have just reviewed evidence suggesting that they also participate in the control of diameter growth. Thus the production of these two hormones could be the basis for the coordination of height growth and diameter growth.

Before we leave the topic of extension and diameter growth it is important to mention the possible role of cytokinins. You will recall that these compounds were discovered as a result of their ability to cause cell division in callus cultures. Cell division also occurs in the growth of height and diameter that we have just examined. Do cytokinins have a role here? The answer is that we have no evidence for it. Cytokinin treatment does not stimulate cell division in shoot apices or in vascular cambia. Indeed, in some species it results in reduced growth. We noted earlier that it is possible to demonstrate an action of a hormone only when there is a deficiency of that hormone. The inability to demonstrate the stimulation of cell division in shoot apices and vascular cambia by CKs is interpreted as suggesting that the tissues already produce ample amounts of them. By analogy with our discussion of the role of GAs in SAM activity, we could attempt to generate a deficiency by the use of anti-cytokinins. For various reasons, however, anti-cytokinins are not as freely available as anti-gibberellins and we are aware of no reports in the literature where they have been used for this purpose. However, a new generation of anti-cytokinins was reported at an international conference in 1991 so this situation may soon be rectified.

We mentioned earlier that monocotyledons do not produce vascular cambia, but they do show diameter growth. This is brought about by the formation of additional vascular bundles from parenchyma cells which gradually push apart the cells already present, thus causing the diameter to increase.

8.5 Apical dominance revisited

We discussed the phenomenon of apical dominance in Chapter 1. Apical dominance refers to the inhibitory effect of the shoot apex on the outgrowth of the axillary buds, laid down by the action of the eumeristem. Thus, the apical bud grows more vigorously than the axillary buds despite the latter being closer to the sources of organic and inorganic nutrients. Since decapitation of the shoot (ie the removal of the apical bud) stimulates the outgrowth of the axillary buds it is obvious that the apex is the source of an inhibitory compound. This was corroborated by experiments which showed that compounds which diffuse out of apices into agar can re-impose inhibition of outgrowth if applied to the cut surface of the shoot. A number of plant hormones have been tested for their ability to mimic the action of the apex (Figure 8.26), but only IAA was able to do so.

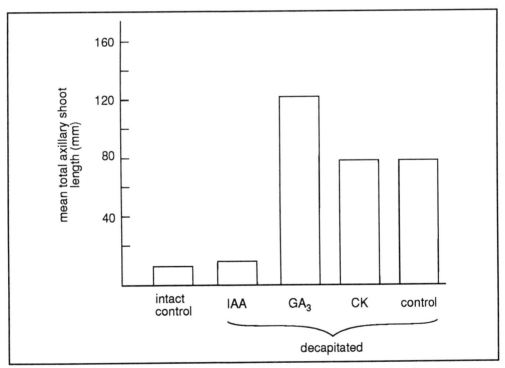

Figure 8.26 Effect of hormones on outgrowth of axillary buds of pea plants after decapitation. Hormones $(1g\ l^{-1})$ were applied in lanolin paste and the outgrowth measured 10 days after decapitation (data from P.F. Wareing and I.D.J. Phillips in Growth and Differentiation in plants, 3rd edition, Pergamon p134, 1981).

The view that IAA is involved in the apical dominance is corroborated by the fact that the auxin transport inhibitor TIBA promotes the outgrowth of axillary buds below the point of its application.

Bearing in mind the dose response curve of auxin activity, shown in Figure 6.3, it was not surprising that the above results led to the suggestion that IAA, generated in the apical bud, entered the axillary buds and accumulated to an inhibitory level. As the stem continued to grow, an individual axillary bud would become farther and farther away from the source of IAA, and its IAA content would gradually fall due to the action of IAA oxidases. Thus, the axillary bud would be gradually released from inhibition and start to grow. However, inhibited axillary buds have been shown to have very low IAA content rather than very high concentrations, so this hypothesis of a direct action of AA does not stand up to close scrutiny.

CK and IAA are required for sustained lateral bud development

Application of CK to inhibited buds caused the temporary outgrowth of the bud. Continued growth was maintained only when IAA was applied with the CK, despite the fact that IAA applied alone had no activity. Cytological analysis showed that CK with IAA caused a permanent increase in cell division activity in the meristem but CK alone changed this only temporarily. This is reminiscent of the actions of the two hormones on cell division in callus cultures. We suggested in Chapter 6 that a major site of CK biosynthesis is the root and it is possible that the role of apex-produced IAA is to direct the transport of root-produced CK to the apex, thus causing a deficiency in the axillary buds. As the stem continues to grow an individual bud finds itself outside the field of influence of the apex and can now receive CK from the root. These ideas are illustrated in Figure 8.27.

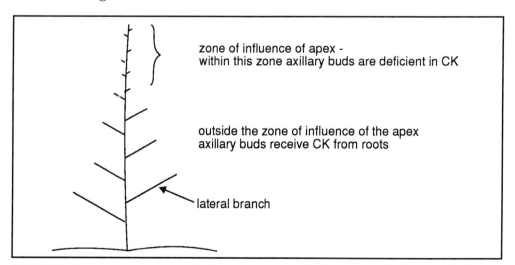

zone of influence of apex -
within this zone axillary buds are deficient in CK

outside the zone of influence of the apex
axillary buds receive CK from roots

lateral branch

Figure 8.27   Diagram to show release of apical dominance in buds which are outside the zone of influence of the apical bud.

This hypothesis has not been tested directly. Even if it is correct, we would still be faced with the problem of how CK could activate axillary buds in the natural situation, whereas CK plus IAA are needed for this in the experimental situation. Thus, we still do not have a full understanding of the phenomenon.

8.6 Auxins and CKs are also involved with shoot and root formation

CKs were discovered as compounds which, in association with an auxin, caused cell division in callus cultures. The original discovery also showed that the two hormones could interact to cause the formation of organised meristems. It was shown that if the proportions of auxin and CK were altered then the type of meristem formed was altered. When the proportion of auxin to cytokinin was relatively high some callus cells gave rise to root primordia which grew down into the medium. A higher proportion of cytokinin to auxin resulted in the formation of shoot primordia which grew upwards from the surface of the callus. This phenomenon has been confirmed with a very wide range of species. If the shoot buds are excised and inserted into a medium containing a high ratio of auxin to CK, roots form at the base of the bud and a complete plantlet has been obtained. It is possible to produce large numbers of such plantlets in a relatively short period of time and this phenomenon is the basis for the new industry of plant production by tissue culture methods. These matters are covered more fully in the BIOTOL texts, 'In Vitro Cultivation of Plant Cells' and 'Biotechnological Innovations in Crop Improvement'.

type of meristem influenced by auxin: CK ratio

8.7 Back to the beginning - phototropism

The first plant hormone to be discovered was IAA, as a result of studies of the phenomenon of phototropism (Figure 6.6). Despite the fact that the phenomenon has been studied for many years and has perhaps received more attention from plant physiologists than any other single topic, controversy still rages. In a recent issue of a major plant research journal, 15 scientists were asked for their views, but no consensus emerged (Plant, Cell and Environment (1992) *15*, 762-794). We will not enter the fray, but will give a somewhat simplified overview.

8.7.1 How is an asymmetry of IAA generated?

The basic hypothesis to explain why plants grow towards the light, put forward by Cholodny and Went, holds that the shaded side of the shoot has a higher IAA concentration than the light-exposed side and this causes more growth on the shaded than the exposed side. An asymmetric distribution of IAA could be generated in two ways. We know that IAA is a photolabile molecule. Perhaps IAA is broken down by light on the exposed side thus producing a gradient across the shoot apex. Alternately, IAA could simply migrate away from the light and travel down the shaded side. Figure 8.28 shows the results of experiments designed to test these two possibilities. In these experiments coleoptile tips are removed, placed on agar blocks and the IAA content of the block subsequently determined.

SAQ 8.14	Which of the two hypothesises is supported by the data of Figure 8.28?

Figure 8.28 f) shows that an IAA gradient of approximately 1:2 is set up across the tip.

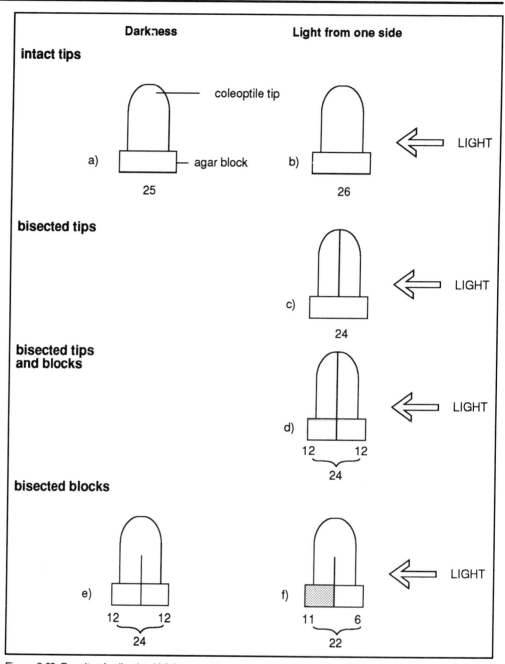

Figure 8.28 Results of collecting IAA from maize coleoptiles treated in various ways. Figures beneath the diagrams indicate the relative amounts of IAA.

Π Examine Figure 6.3 and, assay the linear part of the graph,and calculate the effect on growth of a 2.0 fold increase in IAA.

It would cause a growth increase of approximately 10 per cent above the control.

This is not a large difference in growth rate. Nevertheless, it would cause a gradual bending of the stem towards the light. Thus this evidence is in support of the Cholodny-Went hypothesis. However, we can easily measure the growth rate on the shaded and exposed sides to see how quickly they react. The results of such measurements are shown in Figure 8.29.

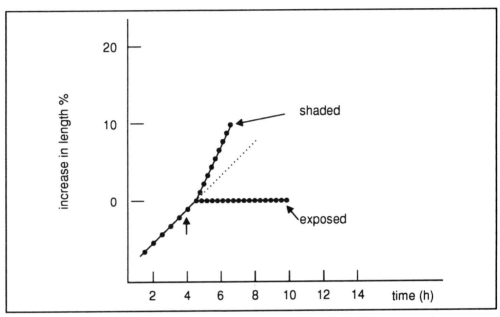

Figure 8.29 Growth of both sides of an *Avena* coleoptile exposed to unilateral light, expressed as a percentage of initial length. The arrow indicates when the unilateral light was switched on.

light-stimulated production of an inhibitor

These results show that whereas the shaded side does show an increase in growth rate, the exposed side virtually stops growing immediately. Results of a similar nature have been obtained with tissues other than coleoptiles including hypocotyls and epicotyls of dicotyledonous species. Further, the examination of the growth rate in zones progressively farther and farther away from the apex shows that this virtual cessation of growth occurs instantaneously in each zone. This phenomenon of the growth cessation is not what would be expected if a growth-limiting compound was being produced in the apex and transported in a basipetal direction. Rather, it suggests the light-stimulated production of an inhibitory compound. We have already mentioned one such inhibitor, raphanusamide (Figure 6.15), which is produced in radish hypocotyls. A recent report shows that a compound with similar activity has been identified in maize shoots as 6-methoxy-2-benzoxazolinone. Both compounds inhibit IAA-stimulated growth but the inhibition is counteracted by additional auxin, suggesting that they are both anti-auxins. There is some evidence showing that unilateral light causes an asymmetric production of such inhibitory compounds. If this is a widespread phenomenon it could be suggested that phototropism is caused by a combination of increased IAA on the shaded side and a reduced *effective* IAA concentration on the exposed side. In other words IAA, as an example of an auxin, is still a central factor. Results of further experiments examining the distribution of light-induced inhibitors are awaited with eager anticipation.

Summary and objectives

Plant growth occurs in meristems and these can be classified as either primary or secondary. Primary meristems control growth in height or length, whereas secondary meristems control growth in diameter. GAs and auxins interact synergistically in the control of both primary and secondary growth. They affect cell elongation by altering wall plasticity but, although much is known about cell wall structure, we still do not fully understand the mode of action of the hormones. Cell growth is also affected by the phytochrome photoequilibrium and here, too, changes in plasticity are involved. This mechanism enables a plant to detect shade. Auxins and cytokinins are involved with the control of apical dominance in shoots and in root and shoot formation in calluses. Auxins were discovered by studying phototropism and although inhibitors are now implicated in this process, they appear to be natural anti-auxins, thus re-emphasizing the central role of auxins.

Now you have finished this chapter you should be able to:

*   show an understanding of the role of meristems in plant growth and interpret data related to the affects of hormones;

*   show an understanding of a growth rate equation linking hydraulic conductivity, water potential, wall yield stress and wall extensibility and interpret data on the effects of IAA and GAs;

*   explain how phytochrome photoequilibria affect plant growth rate and how this activity is accounted for in the growth rate equation;

*   describe phototropism and interpret data to test hypotheses proposed to explain it.

Responses to SAQs

Responses to Chapter 1 SAQs

1.1   Plant a shows a strong apical dominance.

This is evident because side branches even in the older leaves are still quite short and in young leaves the axillary buds are still inactive.

Plant b shows weak apical dominance because bud activity is evident in young leaves and has been proceeding for some time in the older ones.

Note, then, that a shoot which shows strong apical dominance will tend to produce a tall main stem with little branching whereas bushy shoots are produced by weak apical dominance.

1.2   a) matches with 3) and 7).

b) matches with 8).

c) matches with 3) and 4).

d) matches with 2) and 8) (since the torus pit is a pit).

e) matches with 4) and 6).

f) matches with 1) and 5).

1.3   a) matches with 3) and 4).

b) matches with 1), 2) and 5).

c) matches with 1), 6) and 7).

1.4   Wood depends upon xylem tracheids and vessels, and linen depends upon sclerenchyma fibres, both of which are lignified. These cells are dead when mature and so do not carry out respiration.

1.5   a) matches with 2), 4), 5) and 6).

b) matches with 3), 5) and 6).

c) matches with 5).

d) matches with 1).

e) matches with 2), 4) and 6). You may also include 5).

f) matches with 7).

1.6   There is not a simple answer to this question because it depends upon a number of factors. In its simplest form the answer would be yes, as long as the protein was not too big to pass through the gaps between the cell wall polymers. When we have discussed wall structure in detail in Chapter 8 you will be able to give a much expanded answer.

Responses to Chapter 2 SAQs

2.1

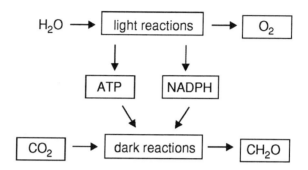

2.2   1) 680nm light would activate PSII thus reducing Phae. The electron would be passed from there to PC, where it would stop because PSI could not accept it. PSII chlorophyll would receive an electron from the splitting of water and so PSII would again be activated. Now, the electron would travel only as far as the cyt b/f complex because PC could not accept it. As PSII continued to 'fire' repeatedly, the earlier intermediary carriers would become reduced further and further back until they were all reduced. Thus, 680nm light would leave the cyt b/f complex reduced.

2) The opposite would happen with 700nm light. The firing of PSI would drag electrons from PC and the cyt b/f complex, but these would not be renewed from PSII because it was not firing. Thus, the cyt b/f complex would be in the oxidised state.

2.3   1) Incubation at pH 4.0 for several hours would be expected to lead to equilibration with the lumen pH. Transfer to pH 8 medium would generate a protonmotive force and addition of ADP and Pi showed that this could be used to generate ATP.

2) When pH 4.0 equilibrated thylakoids were transferred to pH 8 medium containing the hydrogen carrier the protonmotive force would be partially dissipated. This would reduce ATP formation.

3) The 10nm particles seen floating in the suspension could have been CF_1 complexes dislodged by the rough treatment. If so, the CF_0 complex would simply allow protons to diffuse out down their concentration gradient, thus significantly reducing the protonmotive force. This would reduce ATP formation in the remaining intact CF_0/CF_1 complexes.

2.4   1) Although there is enough energy in blue light to drive two photoacts the manner of absorption of the energy will not permit it. Thus, chlorophyll will absorb blue light and be activated to the second or higher excited states, but then immediately release energy as it falls back to the first excited state. This energy is released as heat. If it was released as light it could conceivably be reabsorbed by another pigment molecule, but there is no evidence to suggest that this happens.

2) Energy from 660nm light would be expected to be channelled to both PSI and PSII because the wavelengths and energy levels are suitable. 690nm light, however, has an energy level such that it can only be transferred to PSI.

3) In the presence of excess NADPH the rate of transfer of electrons would be depressed and all the components shown in Figure 2.3 would accumulate in the reduced state. This would delay the transfer of energy from one antenna pigment to the next and the some of the energy may be released as light. If so this light would be at a higher wavelength than 600nm because some of the energy would be lost in transfer. Such light emission is called fluorescence.

2.5   The problem could be solved by the plant carrying out cyclic as well as non-cyclic phosphorylation reactions. We saw earlier that 8 quanta of light produce 1 molecule of O_2 and 2NADPH with 8 protons released into the lumen. If 3 protons produce 1ATP then 8 protons above would not quite manage to produce the 3ATP required. Two photoacts using the cyclic scheme could release two protons from PQH_2 which, together with the 8 from the non-cyclic scheme would produce 10, more than enough for the production of 3ATP.

2.6   The scheme in Figure 2.9 shows how amylose may be made but it does not explain how amylopectin, the branched polymer is made. We give a few more details of how this is achieved in the text following the SAQ.

2.7   The first piece of evidence is very simple. Electron micrographs reveal starch grains in the chloroplast but not in the cytosol.

Secondly, cell fractionation studies confirm this and also show that the enzymes which catalyse its synthesis are found only in chloroplasts.

2.8   **Hint**

Consider the action of the Pi-triose phosphate translocator. What is the consequence of the formation of mannose 6- phosphate?

Response

The formation of mannose 6-phosphate locks up a considerable amount of available Pi. Thus, there is a reduced amount of Pi available to exchange with triose-phosphate via the translocator. Consequently, triose phosphate is retained within the chloroplast, increasing starch, but decreasing sucrose, production.

2.9   Essentially, malate is generated in the mesophyll cells and broken down in the bundle sheath cells, the two cell types being part of the leaf symplasm. This generates a concentration gradient between the two and malate diffuses down this gradient from the mesophyll into the bundle sheath cell. Similarly, pyruvate is generated in the bundle sheath cell and removed in the mesophyll. Thus, its concentration gradient is in the opposite direction and it diffuses out of the sheath cell to the mesophyll.

2.10

$\begin{array}{ccl}
\text{①} & = & \text{assimilation} \\
\text{②} & = & \text{transport} \\
\text{③} & = & \text{decarboxylation} \\
\text{④} & = & \text{transport} \\
\text{⑤} & = & \text{regeneration}
\end{array}$

2.11 There is no single, precise answer to this SAQ but any answer should include some of the points made below.

The C2 PCO cycle is considered to be a scavenging process in which the plant recovers three quarters of the carbons which are converted to 2-phosphoglycolate. If a proposed pathway consumes more energy than it has the potential to release, it will take place only if the pathway generates a vital metabolic component; the cost of the process is said to be justified by the value of the product. The C2 PCO cycle might possibly be seen as a means of generating glycine, serine or glycerate but this is not the 'normal' way in which these compounds are made. Thus, the production of 3-phosphoglycerate can be said to be the only valuable product of the cycle. The 'potential yield' of ATP would be significantly reduced if an ATP molecule was sacrificed at each transfer step between organelles. Thus, it would not be surprising if passive exchange systems operated in the transport of the metabolites between the organelles. These matters are the subject of active research at present.

2.12 The C4 plant would reduce the CO_2 concentration to below the compensation point of the C3 plant. At this CO_2 level the C3 plant would show net loss of CO_2 and would gradually use up its reserves until it starved to death.

2.13 The effects of light and CO_2 can be rationalised by the suggestion that the mechanisms operate to minimise water loss whilst allowing photosynthesis to proceed as actively as possible within the confines of the water loss phenomenon. Thus, because photosynthesis does not proceed in the dark, a mechanism which results in dark closure of stomata will conserve water while not limiting photosynthesis. In the light, stomata should be open to allow CO_2 to enter, unless there are already high internal CO_2 levels. This rationalises the effects of light and CO_2.

Responses to chapter 3 SAQs

3.1   In a perfect vacuum the absolute pressure would be 0 MPa and the P value - 0.1013 MPa. Pure water under positive pressure would have a positive ψ.

3.2 $\psi_{cell} = -0.2 + 0 = -0.2$ MPa

$\psi_{solution} = -0.12$ MPa.

Therefore, water moves from the solution into the cell.

At equilibrium $\psi_{cell} = \psi_{sol} = -0.12$ MPa: but $\psi_{cell} = \pi_{cell} + P_{cell}$.

Therefore $P_{cell} = -0.12 - (-0.2)$ MPa $= +0.08$ MPa.

3.3 ψ_{cell} is now - 0.12 MPa, $\psi_{solution}$ is now - 0.10 MPa.

Therefore, water again enters the cell until ψ_{cell} is - 0.10 MPa.

At equilibrium:

$P_{cell} = -0.10 - (-0.20)$ MPa $= +0.10$ MPa.

3.4   Spaces 1-5 would be occupied by the bathing solution because the cell walls offer no resistance to the movement of small molecular weight solutes.

Unusual shapes, of the type shown at 1, could possibly indicate a particular distribution of plasmodesmata or the location of damaged ones.

3.5   Calculation of the mean change in mass for each chip group gives the following results:

1)   + 0.27g.

2)   + 0.20g.

3)   + 0.01g.

4)   - 0.06g.

5)   - 0.18g.

Thus, chips in solutions 1 and 2 gained mass, chips in solutions 4 and 5 lost mass, but those in solution 3 showed virtually no change in mass. If a tissue absorbs water by osmosis the mass of the tissue will increase due to the mass of the water. Conversely, if a tissue loses water by osmosis its mass will go down. However, if a tissue is placed in a solution of equal ψ to itself there will be no net movement and, therefore, no change in mass.

The above results shows that the ψ of the potato tissue is very close to that of the 0.20 mol l^{-1} sucrose solution. Solutions 1 and 2 have ψ values greater than that of the potato and solutions 4 and 5 ψ have values less than that of the potato.

3.6   Any system of measurement involves inaccuracies, and errors are additive. Calculation of P from ψ and π sums the inaccuracy. Thus, the P value obtained in this way is not as reliable as one would wish.

3.7   The description given in the text of placing a severed twig in a pressure chamber and applying pressure until solution is exuded from the cut surface not only shows that the xylem fluid is under tension but also allows its magnitude to be measured. The pressure needed to bring the solution back to the surface is equal in magnitude but opposite in sign to the negative pressure that existed in the xylem before the twig was severed.

3.8   1) Sensor number 1 at each unit will detect a temperature change first, signifying an upward sap flow and flow would be detected first at unit A, then B and finally C.

2) Sensor number 1 will again detect flow first but the order will be C,B,A.

3.9   Cuttings do not have roots, so the experiments are being conducted under rather abnormal conditions

3.10   1) Replacing water by sucrose would reduce the difference in ψ between the solution and the leaf and this would reduce the rate of transpiration.

2) Removal of roots would remove the root resistance, thus the rate would be expected to increase.

3) Excising roots in air would allow air to enter the vessels and tracheids thus blocking them. The rate of transpiration would be expected to fall.

4) The sorts of morphological features you might find in a leaf which would indicate that the leaf was specialised to reduce water loss should include:

- a thick cuticle;
- sunken stomata;
- stomata at some distance from conducting vessels;
- large number of cell layers in the mesophyll with very few air spaces between the cells;
- surface hair.

Leaves specialised to reduce water loss may have combinations of these features.

3.11   The demonstration of 2, 4 and 5 would be necessary. If ABA has been found in the epidermal apoplast it must come from somewhere. The presence of high ABA levels in the mesophyll would negate the idea of a root signal so alternative 3 must be wrong and alternative 4 correct. If the leaves show wilting (alternative 1), this would suggest that the root was unable to cause stomatal closure before the leaf experienced water stress. Thus 1) is not consistent with a root-to-shoot signalling system.

Responses to chapter 4 SAQs

4.1   Aeration provides oxygen to the roots. In the absence of aeration, oxygen becomes depleted in the solution and this reduces the respiration rate of the roots. Such roots will show slow growth and the whole plant will be stunted as a result.

4.2   The salts used by Knop and by Sachs, although the best available at the time, were impure and contained many other elements as contaminants. Table 4.1 shows that if molybdenum is present as a 0.0001% impurity in the sample of nitrate its requirement will be met and the essentiality missed. Thus the full list became known only when very pure chemicals became available.

4.3   Potassium and boron were omitted.

4.4   Since each chlorophyll molecule requires an atom of magnesium, a deficiency of this element will lead to defective chlorophyll production and the leaves will be pale yellow in colour. This condition is called chlorosis.

4.5   The apoplast of the root hair is directly in contact with the soil.

4.6   The 'easy' one is the selective action of the plasmalemma of the stelar parenchyma. This membrane will exhibit control over what is transported to the xylem and subsequently to the shoot.

The second one is more tricky. The hint is to think of the pathway taken during the radially-inward movement of ions and also the definition of the symplasm. The answer is absorption into the vacuole. Any ion transported into the vacuole is taken out of the diffusion stream and will be held back.

4.7   The Casparian strip prevents outward diffusion in the same way it prevents inward diffusion.

4.8   Selection takes place at (1) entry to the root symplasm, (2) entry to the transpiration stream in the xylem and (3) entry to the leaf symplasm.

Selection at the entry to the transpiration stream contributes to the ability of the plant to maintain a mineral regime in the shoot which is different from that in the root.

Selection at the entry to the leaf symplasm allows the specialised cells of the leaf to absorb the minerals they require and reject those they do not need.

4.9   This is an interesting question to which there will be a variety of answers. It depends on how you define essential. If you decide that the plants natural environment should be included when assessing essentiality you most likely conclude that sodium is essential. You have to think carefully, however, about how you could demonstrate it. It would be no use growing the plant in sea water because you cannot generate a minus sodium control. If we rephrase the question you may be able to see an answer. How can it be demonstrated that sodium is required by *Spartina* to allow it to survive in solutions with a very negative π value? This could be done in two different ways.

1) We could provide a concentrated mineral solution by using elements other than sodium, supply sodium at low concentration and see if the plant survives. This might work but we would run into complications if any of the minerals used were toxic at high concentrations.

2) The alternative is to provide a normal mineral mixture but reduce π further by adding a compound which is not absorbed. The compound polyethylene glycol has been widely used for such purposes. *Spartina maritima* can be shown to require sodium under these conditions.

4.10   Your answer will take a variety of forms but the points below should be covered in one way or another.

The concentration of K^+ and Ca^{2+} ions in the xylem exudate is considerably higher than in the external solution suggesting an active energy-requiring absorption and secretion process. Na^+ ions are only marginally more concentrated so no clear conclusion about its absorption and secretion can be drawn. K^+ and Ca^{2+} in the guttation fluid are at a much lower concentration than in the xylem exudate. This suggests a strong absorption by the leaves. In contrast, Na^+ ions are at a higher concentration in the guttation fluid implying that the leaves strongly discriminate against sodium and secrete it rather than absorb it into their cells.

4.11   The stripped plant shows considerably higher potassium levels in the xylem than in the phloem, in regions where they are separated. This suggests that the xylem is the site of transport of potassium. Sections 1 and 8 in the stripped plant and all sections in the unstripped plant show significant potassium concentrations in the phloem. This would suggest that potassium can move laterally from the xylem into the phloem.

4.12

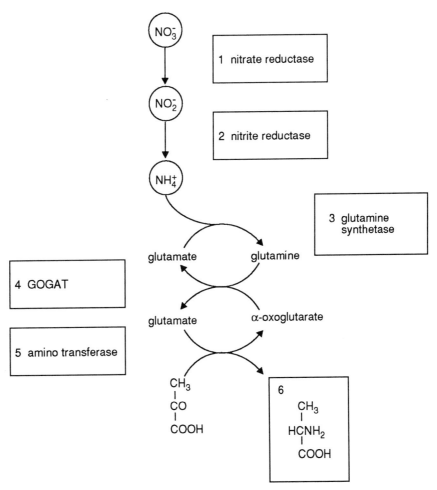

4.13   The proportional increases in nutrient absorption are N 50%, P 135%, K 46%. These can be calculated using the relationship:

proportional increase (as %) =

$$\frac{\text{(content with mycorrhiza)} - \text{(content without mycorrhiza)}}{\text{content without mycorrhiza}} \times 100\%$$

Thus phosphorus absorption is very markedly stimulated as might be expected but so, too, is nitrogen and potassium absorption. It could be suggested that growth in the absence of mycorrhiza was phosphate limited. Once this limitation was removed, growth and also N and K absorption was greatly stimulated. Thus, it is proposed that the N and K effects are indirect ones, but these results show that great care must be used when interpreting such experiments.

4.14   The relationship between the binding of Zn to the plant cell walls and the plants' tolerances of Zn is remarkably direct in these strains of *Agrostis tenuis*. It shows that strains of this species have a variable ability to bind Zn in their walls and that this could be preventing Zn absorption. Thus this species appears to achieve tolerance by preventing Zn gaining access to the cytoplasm.

Responses to chapter 5 SAQs

5.1   *Hint:* Xylem cells are dead and phloem cells alive. Thus, steam ringing will kill the phloem, and any other living cells in the zone, but it should not affect the xylem.

The treated leaf retained turgidity showing that the transpiration stream had not been interrupted. Thus, the xylem was able to transport water. The distribution of ^{32}P was the same in the treated and control plants suggesting that the xylem was also able to transport minerals. However, the steam ring blocked the movement of ^{14}C suggesting that it was being transported in the phloem.

5.2   The main risk with an experiment of this type is that the treatment might be actually damaging the tissue thus giving artificial results, termed artifacts. One check would be to return the pressure to zero after the 1.05 MPa treatment and see if the apical bud recovered its ability to import radiolabel. Having shown such a recovery, it would be of interest to find out what pressure would be needed to damage the bud and render it unable to recover. If this pressure was significantly higher than that from which recovery could be demonstrated, this would give strength to the supposition that the results of lower pressures were valid.

5.3   In the light no stem curvature occurred unless 2,4-D was applied to the leaf. This suggests that 2,4-D was translocated in the light. The data suggest that no 2,4-D was translocated in the dark unless sucrose was also present. The darkened leaf would not be photosynthesising and under this condition no 2,4-D was translocated. The applied sucrose would take the place of sucrose generated in photosynthesis and could allow pressure gradients to be generated. This experiment also shows that 'extraneous' substances, such as 2,4-D, can be transported and is an argument in favour of mass flow. Even such things as viruses have been found to be translocated in the phloem.

5.4   Any zone of the plant which utilises sucrose, but does not generate it itself, will be a sink. Thus, developing or immature seeds and fruits are sinks as are the other storage organs of plants. These include tubers, rhizomes, bulbs, and the storage roots of such plants as carrot, parsnip, beetroot, etc. Flowers are also sinks since, in most cases, they do not contain chloroplasts. As we saw above, the apical bud is also a sink. All areas of active growth may be regarded as sinks, including immature leaves.

5.5   PCMBS should have no effect. It would be expected to inhibit sucrose absorption from the apoplasm into the symplasm but since transport is symplastic there is nothing for PCMBS to inhibit.

Plasmolysis would be expected to reduce sucrose transport because it breaks some of the plasmodesmata which link adjacent cells.

5.6   You could make use of the inhibitor, PCMBS, which inhibits the activity of the sucrose-proton symport. Treatment of excised young seeds with radiolabelled sucrose would be expected to result in the young seed absorbing sucrose and becoming radioactive. If PCMBS inhibits the process it would suggest the involvement of the symport.

It could also be demonstrated by treatment with compounds which dissipate a proton gradient, such as dinitrophenol.

5.7   This would require the stripping open of the bark of the stem adjacent to the leaf and separating the xylem and phloem by waxed paper prior to applying the radiolabel. After a suitable period, 16h for example, the xylem and phloem tissue would be excised and their radioactivity separately determined. As a corollary, the petiole of the leaf on a duplicate plant could be steam girdled prior to the application of ^{32} PO$_4$. Since the xylem is dead and the phloem alive, the girdle should prevent ^{32}PO$_4$ transport if it occurs in the phloem but not if it occurred in the xylem.

Responses to chapter 6 SAQs

6.1   You could see if they stimulate the growth of coleoptiles in the oat coleoptile bioassay. Degradation products are not active in this test.

6.2   **Hint:** To measure the rate we need to know when radioactivity is first received in the receiver. We can get a good estimate of this by plotting the data and extrapolating the line to the x axis.

Answer: The data are plotted blow, the extrapolated line crosses the x axis at almost 0.8 h. Thus:

$$\text{Transport rate} = 5.44 \text{ mm in } 0.8 \text{ h} = 6.88 \text{ mm h}^{-1}$$

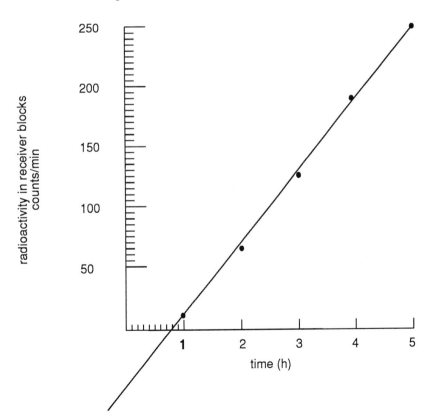

6.3   The results could suggest that IAA is synthesised in the apical part of the shoot and transported in a basipetal direction down the epicotyl and hypocotyl. IAA oxidase in the stem would degrade IAA and the combined affect would be a peak of IAA at the apex and a gradually reducing concentration towards the base.

6.4

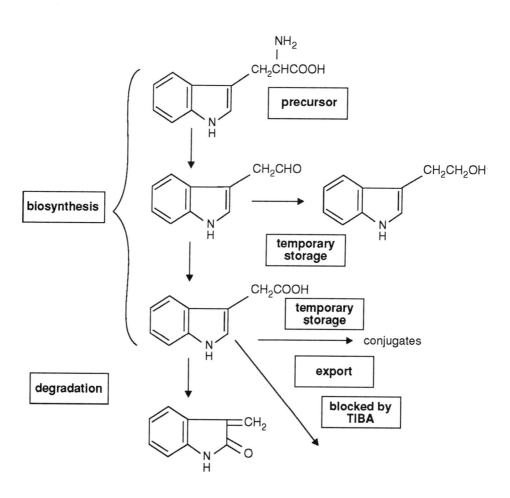

6.5   IAA is photoabile and also broken down by IAA oxidase but synthetic auxins are not. Thus, synthetic auxins are effective at lower concentrations.

6.6   Remember that 1-NAA is a synthetic auxin (Figure 6.14) and its action will be added to that of IAA already present in the tissue. In experiment 1, the action of all 1-NAA concentrations was to reduce growth below that of the control. This could suggest that the control roots were producing an optimal amount of auxin in which case the addition of the synthetic auxin 1-NAA, would raise the auxin concentration to supra-optimal levels, thus causing inhibition of growth, as shown in Figure 6.3.

Figure 6.14 and 6.15 show that 2-NOA is an auxin and 1-NOA an anti-auxin but that 1-NAA is an auxin. By analogy it might be expected that 2-NAA is an anti-auxin. If control roots contain an optimal amount of auxin we might expect an anti-auxin to cause reduced growth. 2-NAA produced this affect in experiment 2. If this interpretation is correct we might expect 1-NAA to compete with 2-NAA and to cause a stimulation of growth compared with the 2-NAA treatment. In experiment 2, both 10^{-5} and 5×10^{-5} mol l^{-1} 1-NAA, when added with 2-NAA, showed such a stimulation and the latter returned the growth to the control value.

6.7
1) Figure 6.17 shows that the terpenoid pathway leads not only to GAs but also to steroids and carotenoids. Assuming that the action of the anti-GAs is restricted to Steps 8, 9 and 10 of the pathway shown in Figure 6.17 we would expect an increase in the synthesis of steroids and carotenoids.

2) Anti-auxins affect the action of auxins but not their synthesis; anti-GAs do the opposite.

3) The statement is not true. Naturally-occurring compounds of both types have been found.

6.8
All plants and parts of plants require minerals so we would need to provide all of the essential minerals described in Chapter 4. Because callus is not photosynthetic we must also provide a source of fixed carbon and energy and sucrose is most often used.

6.9
Cell division requires the chromosome number to be doubled prior to mitosis and that involves DNA synthesis. Cells unable to divide could conceivably lack the ability to manufacture the nucleotides needed for DNA synthesis. Thus, a potential source of these was a logical addition to the medium. Herring-sperm DNA was simply the most readily-available source at the time.

6.10
If degradation is inhibited and synthesis continues, CK levels should increase in the tissue.

6.11
IAA is made from tryptophan and CKs from adenine. These two compounds are very important in general plant biochemistry and mutations which caused the loss of the ability to make them would almost certainly be lethal mutations. Thus, they would not survive for us to study.

6.12
1) GA_1 is active in all bioassays and this is in agreement with it being at the end of this pathway. *Dwarf*-1 is presumably blocked between GA_{20} and GA_1, *dwarf*-z between GA_{19} and GA_{20}, *dwarf*-y between GA_{53} and GA_{19} and *dwarf*-x between GA_{12} aldehyde and GA_{53}. These results are in agreement with the pathway shown in Figure 6.19.

2) If GA_{19} had produced a 100 % response in all mutants it would have suggested that the pathway in Figure 6.19 was incorrect since it would indicate that GA_{19} came after GA_{20} in the pathway.

Responses to chapter 7 SAQs

7.1   Potato matches with 1), 2), 3) and 4).

Strawberry matches with 1), 2) and 4).

Bulbs match with 1) and 2).

Corms match with 3).

Rhizomes match with 1), 2), 3) and 4).

Root budding matches with 4) and 5).

Tuberous root matches with 3) and 4).

7.2   a) = 1), 4).

b) = 1), 4), 7).

c) = 3), 6), 8).

d) = 1), 6), 8).

7.3   Your drawing should be similar to the one below. Each ovule develops into a seed and the mature fruit consists of the pod wall, formed from the pericarp, surrounding the row of mature seeds:

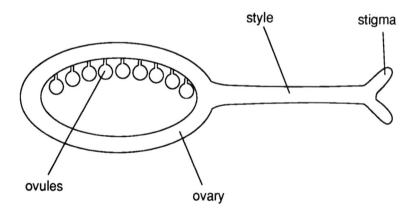

7.4   Dry seed shows very low metabolic activity and did not respond to chilling. This suggests that the response to chilling involves a positive metabolic process. This is corroborated by the effects of the different temperatures during the 85 day exposure. If chilling was causing some sort of physical effect, we might expect that the colder the temperature the more effective it would be. This was clearly not the case, there being an optimum chilling temperature at approximately 5°C.

7.5   Hint: remember that most plants release their seed in autumn. What would happen if they germinated immediately?

The mechanisms of seed dormancy have the effect of ensuring that the seed does not germinate until the spring of the following year. This ensures that the offspring have a full growing season ahead of them in which to become established. If this confers an advantage to the plant then any mechanism which achieves it will be selected for. Plant species show incredible variation in their form and this must reflect differences at the biochemical level. It is not surprising, therefore, to find an array of dormancy mechanisms. Indeed, it would be very strange if there was only a single mechanism.

7.6   1) The critical lengths are 16 hours for *Pharbitis* and 12 hours for *Sinapis*.

2) a) *Xanthium* will not flower because the photoperiod is greater than its critical daylength. *Hyoscyamus* flowers because its critical daylength is exceeded.

b) Both species flower.

c) Only *Xanthium* flowers.

7.7   Treatments 4 and 5 produced the same results as treatments 1 and 2 because interruptions of the day have no effect.

	Soybean	Spinach
1)	V	F
2)	F	V
3)	V	F
4)	V	F
5)	F	V

7.8   This can be done in two ways. The leaves can be covered during the photoinductive exposure and then uncovered. Alternatively, the leaves could be cut off prior to the photoinductive treatment. This is more drastic and does slow down subsequent development, but it demonstrates the same point, that flowers are produced only when the leaves are exposed.

7.9   These experiments are important because they demonstrate the universal nature of the concept of florigen. The universal nature of florigen could be further tested by grafting photo-induced SD and LD donor branches onto day-neutral plants (DNP). Such experiments have, in fact, been done, and they also resulted in flowering.

7.10   The plants will flower under the short days of the autumn. Theoretically, we might expect them to flower under the short days of spring but try to remember what we said about soil temperature at the time of the spring equinox. Soil temperatures are still hardly above 6°C at this time so new shoots will normally not be produced until well into spring. Thus, at the time of the spring short days the plant has hardly started to develop and cannot yet respond to the photoperiod. Thus, it flowers naturally in autumn.

Responses to Chapter 8 SAQs

8.1   The root cap prevents the cell division zone from coming into direct contact with the potentially damaging soil particles. The mucilage acts as a lubricant and helps the root to slide between the soil particles.

8.2   The shoot apex is covered by the overlapping young leaves, which provide protection against mechanical damage and water loss.

8.3   Both meristems show division, enlargement and differentiation. However, the shoot differs in having two division zones and in producing leaves and axillary buds as well as axial material.

8.4   The untreated plant had produced leaves and axial material, and so its eumeristem appeared to be functional. Treatment with GA_3 caused the progressive activation of cell division until it was evident 14 mm below the apex. This is consistent with the SAM being initially inactive and being activated by GA_3.

8.5   **Hint**: think of a way to reduce the amount of GA produced.

        Response: It might be possible to reduce SAM activity by treatment with an anti-GA. The hypothesis could be tested by examining the effect of the anti-GA on the number of mitotic figures in the sub-apical zone. If the anti-GA reduced SAM activity due to a deficiency in GA, treatment with GA should overcome its action and return the number of mitotic figures to normal.

8.6   The reasons are:

-   the shape of the growth response to the two hormones is quite different. There is a distinct time lag before the tissue responds to GA_3, but this is not the case for IAA;

-   IAA and GA_3 applied together produce a synergistic effect (ie their effect is more than additive). This would not happen if they acted on the same part of the system.

8.7   1) The value of $(\psi_{ext} - \psi_{int}) = 0$ and therefore below -Y and so the growth rate would be zero (see also Figure 8.10).

        2) The growth rate would increase.

        3) The growth rate would decrease.

        4) The growth rate would increase.

        5) The growth rate would increase.

8.8   IAA has altered the wall extensibility and GA has altered the wall yield stress.

8.9   The application of high levels of calcium caused a dramatic reduction in wall extensibility whereas removal of calcium ions with EGTA caused a large increase in plasticity, relative to the control.

8.10   Water will move into the cell by osmosis, down the ψ gradient and, in the absence of any other factors directing it, will simply dilute the cytoplasm.

8.11   The stem apex will now be exposed to full light with its red: far-red ratio of about 1.1. Thus, its growth will revert to the slower stem growth and increased leaf growth characteristic of full sunlight.

8.12   IAA is transported in a basipetal direction in stems and an acropetal direction in roots, in what is referred to as the polar transport system.

8.13   The action was synergistic, ie it was more than the added effects of the hormones when applied separately.

8.14   The data support the hypothesis that IAA migrates way from the light:

a) and b) show that similar amounts of IAA are produced in the light and in the dark;

c) shows that bisecting the tip does not affect IAA production;

d) that the IAA reaching separated agar halves is identical if the whole tip is bisected;

e) and f) show that IAA asymmetry is caused by unilateral light only when the agar and the base of the tip is bisected.

Suggestions for further reading

R.M. Devlin and F.H. Witham, Plant Physiology, Willard Grant (1983).
ISBN 0 8715 0765 X

J. Forbes and R. Watson, Plants in Agriculture, Cambridge University Press (1992).
ISBN 0 5214 1755 4

F. Salisbury and C. Ross, Plant Physiology, Wadsworth (1991).
ISBN 0 5349 8390 1

H. Smith, Phytochrome and Photomorphogenesis, McGraw Hill (1984).

L. Taiz and E. Zeiger, Plant Physiology, Benjamin Cummings (1991).
ISBN 0 8053 0245 X

P.F. Wareing and I.D.J. Phillips, Growth and Differentiation in Plants, Pergamon International (1981).
ISBN 0 0802 6350 X

Index

Index